PROFIT

BEE-KEEPING

FOR

SMALL-HOLDERS AND OTHERS

BY

HENRY GEARY, F.E:S.

Expert to the Leicestershire Bee-Keepers' Association, etc. etc.

AUTHOR OF

" **BEES** FOR PROFIT AND PLEASURE ' ETC.

WITH 13 PHOTOGRAPHIC AND OTHER ILLUSTRATIONS

British Library Cataloguing-in-Publication Data
A catalogue record for this book is available from the
British Library

Bee Keeping

Beekeeping (or apiculture, from Latin: *apis* 'bee') is quite simply, the maintenance of honey bee colonies. A beekeeper (or apiarist) keeps bees in order to collect their honey and other products that the hive produces (including beeswax, propolis, pollen, and royal jelly), to pollinate crops, or to produce bees for sale to other beekeepers. A location where bees are kept is called an apiary or 'bee yard.' Depictions of humans collecting honey from wild bees date to 15,000 years ago, and efforts to domesticate them are shown in Egyptian art around 4,500 years ago. Simple hives and smoke were used and honey was stored in jars, some of which were found in the tombs of pharaohs such as Tutankhamun.

The beginnings of 'bee domestication' are uncertain, however early evidence points to the use of hives made of hollow logs, wooden boxes, pottery vessels and woven straw baskets. On the walls of the sun temple of Nyuserre Ini (an ancient Egyptian Pharo) from the Fifth Dynasty, 2422 BCE, workers are depicted blowing smoke into hives as they are removing honeycombs. Inscriptions detailing the production of honey have also been found on the tomb of Pabasa (an Egyptian nobleman) from the Twenty-sixth Dynasty (c. 650 BCE), depicting pouring honey in jars and cylindrical hives. Amazingly though, archaeological finds relating to beekeeping have been discovered at Rehov, a Bronze and Iron Age archaeological site in the Jordan Valley, Israel.

Thirty intact hives, made of straw and unbaked clay, were discovered in the ruins of the city, dating from about 900 BCE. The hives were found in orderly rows, three high, in a manner that could have accommodated around 100 hives, held more than 1 million bees and had a potential annual yield of 500 kilograms of honey and 70 kilograms of beeswax!

It wasn't until the eighteenth century that European understanding of the colonies and biology of bees allowed the construction of the moveable comb hive so that honey could be harvested without destroying the entire colony. In this 'Enlightenment' period, natural philosophers undertook the scientific study of bee colonies and began to understand the complex and hidden world of bee biology. Preeminent among these scientific pioneers were Swammerdam, René Antoine Ferchault de Réaumur, Charles Bonnet and the Swiss scientist Francois Huber. Huber was the most prolific however, regarded as 'the father of modern bee science', and was the first man to prove by observation and experiment that queens are physically inseminated by drones outside the confines of hives, usually a great distance away. Huber built improved glass-walled observation hives and sectional hives that could be opened like the leaves of a book. This allowed inspecting individual wax combs and greatly improved direct observation of hive activity. Although he went blind before he was twenty, Huber employed a secretary, Francois Burnens, to make daily observations, conduct careful experiments, and keep accurate notes for more than twenty years.

Early forms of honey collecting entailed the destruction of the entire colony when the honey was harvested. The wild hive was crudely broken into, using smoke to suppress the bees, the honeycombs were torn out and smashed up — along with the eggs, larvae and honey they contained. The liquid honey from the destroyed brood nest was strained through a sieve or basket. This was destructive and unhygienic, but for hunter-gatherer societies this did not matter, since the honey was generally consumed immediately and there were always more wild colonies to exploit. It took until the nineteenth century to revolutionise this aspect of beekeeping practice – when the American, Lorenzo Lorraine Langstroth made practical use of Huber's earlier discovery that there was a specific spatial measurement between the wax combs, later called *the bee space*, which bees do not block with wax, but keep as a free passage. Having determined this bee space (between 5 and 8 mm, or 1/4 to 3/8"), Langstroth then designed a series of wooden frames within a rectangular hive box, carefully maintaining the correct space between successive frames, and found that the bees would build parallel honeycombs in the box without bonding them to each other or to the hive walls.

Modern day beekeeping has remained relatively unchanged. In terms of keeping practice, the first line of protection and care – is always sound knowledge. Beekeepers are usually well versed in the relevant information; biology, behaviour, nutrition - and also wear protective clothing. Novice beekeepers commonly wear gloves and a hooded suit or hat and veil, but some experienced beekeepers elect not to use gloves because they inhibit delicate manipulations. The face and neck are the most important areas to protect (as a sting here will lead to much more pain and swelling than a sting elsewhere), so most beekeepers wear at least a veil. As an interesting note, protective clothing is generally white, and of a smooth material. This is because it provides the maximum differentiation from the colony's natural predators (bears, skunks, etc.), which tend to be dark-coloured and furry. Most beekeepers also use a 'smoker'—a device designed to generate smoke from the incomplete combustion of various fuels. Smoke calms bees; it initiates a feeding response in anticipation of possible hive abandonment due to fire. Smoke also masks alarm pheromones released by guard bees or when bees are squashed in an inspection. The ensuing confusion creates an opportunity for the beekeeper to open the hive and work without triggering a defensive reaction.

Such practices are generally associated with rural locations, and traditional farming endeavours. However, more recently, urban beekeeping has emerged; an attempt to revert to a less industrialized way of obtaining honey by utilizing small-scale colonies that pollinate urban gardens. Urban apiculture has undergone a renaissance in the first decade of the twenty-first century, and urban beekeeping is seen by many as a growing trend; it has recently been legalized in cities where it was previously banned. Paris, Berlin, London, Tokyo, Melbourne and Washington DC are among beekeeping cities. Some have found that 'city bees' are actually healthier than 'rural bees' because there are fewer pesticides and greater biodiversity. Urban bees may fail to find forage, however, and homeowners can use their landscapes to help feed local bee populations by planting flowers that provide nectar and pollen. As is evident from this short introduction, 'Bee-Keeping' is an incredibly ancient practice. We hope the current reader is inspired by this book to be more 'bee aware', whether that's via planting appropriate flowers, keeping bees or merely appreciating! Enjoy.

CONTENTS

LIST OF PLATES

PROFITABLE
BEE-KEEPING

INTRODUCTION

No small-holder should be without a few stocks
of bees. If he neglects to provide these adjuncts
to his other forms of enterprise he is not utilizing
to their fullest extent the means which lie to his
hand. This may seem a very bold statement, but,
notwithstanding, it is a true one, and one which
is not likely to be contradicted by anyone ac-
quainted with the science of bee-culture.

As is well known, the very foundations of success on small-holdings rest on the tenant deriving
his income from more than one source. In fact,
and within reason, the more irons he has in the fire
the better, providing he thoroughly understands
his various ventures. The small-holder must not
be dependent on any single crop, but must ever
have a reserve of force to counteract any possible
failure. By these means he guards against the
vicissitudes of this most uncertain climate, the
loss of stock by disease, and the falling of markets

owing to the superabundance of any particular kind of produce. Thus his cows and pigs will have support in poultry and bees. These in turn are backed up by fruit and vegetables, and so forth. Working on these lines total loss is practically impossible, whatever the conditions, and the knowledge of this should go far to make the small farmer an optimistic man.

Now among all the kinds of stock or crops which may be worked upon a small-holding, there is not one which will give the returns which bees will give, taking an average of seasons and providing they be properly managed. It is quite a common thing to find apiaries which recouped their owner for his initial outlay during the first season. With what other stock is there even a possible chance of this being done?

Properly worked, bees will show a profit in practically any situation, although the amount of this profit will vary greatly. This is on account of the variation in the honey producing power of different districts. The difference is very great in some cases where special florage is available, taking as an instance apiaries situated within reach of both clover and heather. Apiaries in such districts have been known to show a profit of 50/- per colony in a good season. Speaking generally, however, and taking an average of years, bees should show a profit of 20/- per colony per annum. When this is compared with the average price of a stock of bees, which is about 25/-, and the upkeep about 3/- yearly, further comments on the profitable nature of the pursuit are superfluous.

Again, as regards selling the produce the beekeeper is in a most enviable position. The supply of first-class British honey is not nearly sufficient to meet the demand, and good prices are readily realized. The imports of foreign honey into this country are of the value of about £35,000 annually. This honey finds a market, apart from its use in manufacture, mainly owing to the paucity in the home supply, for which there is an ever increasing demand.

There is no foreign honey which can compare in quality with the native article. A further point is that good honey will keep for an indefinite period without deteriorating in any way, and, should low prices rule, the bee-keeper can hold his produce for a better market without suffering loss. This necessity does not often arise, except in the case of a honey glut, as in the record year of 1906. Many poor colonies secured fifty pounds of surplus honey in that year, and some of my own yielded over a hundredweight each in a medium district.

It will thus be seen what an aid to success a well-ordered apiary can be, and it is within the power of nearly every man to become a successful bee-keeper. The attention required is really very little. No special location is necessary, as any rough corner which is not suitable for cultivation will generally answer admirably, for the bee-hives.

The main thing is to understand thoroughly the habits of the bees and the working of the hive. It is with the intention of teaching this that this little book has been written. As regards the text, abstruse wording has been carefully avoided, and the idea has been to produce a plain, straight-forward work for the use of the great body of small-holders. The number of necessary appliances has been kept within strict limits. Many bee-keeping appliances which figure in catalogues can readily be dispensed with, and thus a certain amount of capital is retained.

As in every other pursuit much money may be uselessly spent, and my object has been to elimi-nate all articles which may be dispensed with without injuring in any way the efficiency of the practical work.

By the same rule the anatomy of the bee and the economy of the hive have only been touched upon sufficiently to serve the everyday needs of the apiarist.

CHAPTER I

THE ECONOMY OF THE HIVE

BEFORE commencing an explanation of the practical operations connected with bee culture, it will be well to devote a few lines to a brief description of the bee with which we have to deal, and which is indigenous to these islands.

The honey-bee is classed by entomologists as follows: Class, *Insecta*; Order, *Hymenoptera*; Family, *Apidæ*; Genera, *Apis*; Species, *Mellifica*; and finally the various varieties—English, Carniolan, Italian, etc., as the case may be.

The honey-bee is possessed of six legs—anterior, intermediate, and posterior. The posterior legs in the case of workers are fringed with stiff bristles, forming the well-known pollen baskets, in which the pollen is conveyed to the hive. They have two pairs of membranous wings, while the framework of the body consists of an external skeleton composed of a horny substance known as chitine, arranged in the form of segments in the abdomen, each segment being formed by a dorsal and ventral plate. The whole body is more or less thickly covered with hair. There are three distinct kinds of bee in a hive, all of which have much in common, but differ in many important particulars. The drone or male bee is stingless, and is also destitute of pollen baskets; while the queen, the only fully developed female in the hive, possesses a sting, and is also endowed with a series of productive organs. She alone of all the inmates of the hive can perpetuate the race. The worker bee is an undeveloped female, and it is this bee which alone performs the whole of the work in the bee kingdom. She it is who gathers the honey, pollen, propolis and water, feeds the young larvæ, builds the combs and protects the colony from attack, finally dying in harness.

A good queen will lay from two to three thousand eggs per day during the height of the breeding season, and she is usually at her best in her second year. After this time she gradually fails, and should be supplanted by a younger mother bee. If left alone the bees will often do this for themselves, but the careful apiarist leaves nothing to chance, and elects to do it for them as a rule. The queen is the centre round which the whole prosperity of the colony revolves, and without good young queens the best results cannot be obtained. She is the mother of the whole of the other inmates of the hive, and has the marvellous ability of laying eggs which will produce at will either drones or worker bees. The queen is absolutely the same as a worker bee at birth, but is reared in a special cell and is fed with special food, and it is this food alone which is sup-

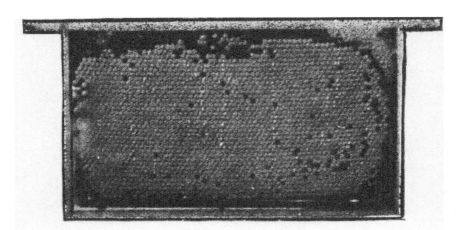

WORKER BROOD

A fine Comb containing practically no drone cells

INTERIOR OF A STRAW SKEP

Note the irregular formation of natural Combs

posed to bring about the evolution. The worker bees being debarred from the stimulating food which conduces to the perfection of the queen are rendered physically incapable of mating with the drone, and therefore can never head a colony. The worker bee can lay eggs, and does so at times, but these eggs produce drones only.

This curious feature of reproduction without fecundation is known as parthenogenesis.

The eggs laid by the queen hatch on the third day, and after passing through the larval and chrysalis stages peculiar to insects, the fully-developed bee hatches on the fifteenth or sixteenth day if it be a queen, on the twenty-first day if a worker, and on the twenty-fourth day if it be a drone. These dates are taken from the time the egg is laid.

A short survey has now been given of the life-history of the bee and of its anatomy. This account is sufficiently full for inclusion in a practical manual, and now we will take the course of events which have place in a normal colony during a season's working. By this means the following chapters will be readily understood and easily put into practice.

Towards the end of February bees begin to move about more freely, and to shake off the lethargy of their winter semi-hibernation. The queen will begin to lay eggs, a tiny circle at first in the centre of the cluster, which rapidly enlarges as the days grow longer and pollen

B

becomes more abundant. The food supply, which has been reduced but little during the winter months, now dwindles rapidly with so many young mouths to feed, and the population of the hive begins to show a marked increase. This increase goes steadily on, and with the first fruits of the new season's honey, gathered from the willows and fruit bloom, breeding quickly rises towards its height. The hive now becomes very congested, and drone eggs will be laid. Towards the end of April drones will be hatching out in forward colonies, showing that the season is commencing in earnest. With May in, good colonies will be packed to the verge of suffocation, and queen-cells will be built in which the queen will lay eggs, and when these are seen it is a sign that the bees are thinking of swarming.

If adverse weather comes on these cells may be pulled down, but otherwise if they are completed a swarm may be expected at the capping of the first cell, providing the weather is suitable.

This swarm will be headed by the old queen. On the ninth day after the first swarm leaves the hive a second swarm or cast will probably be thrown off, headed by the young queen first hatched, and if no restraint has been put upon the bees other small after-swarms may issue. These after-swarms are headed by unfertile queens, which, after hiving, fly forth to meet the drone. And it should be thoroughly understood that a queen only leaves the hive for two purposes, either to lead a swarm or to meet the drone for the act of fertilization.

After the honey season breeding gradually slows down in the colonies, so that by the end of August very little brood is to be found in the hives, unless there is a late flow of nectar. The bees now begin to think of winter, all stores are sealed up, and the hive is made practically air-tight, as regards the top and sides, by the use of a resinous substance called propolis, gathered from the limbs and branches of trees. The drones are usually killed off by the end of July, there being no further need of their services, and incidentally it may be noted that the presence of drones in a hive during the autumn or winter months is almost a sure sign of queenlessness. Towards the autumn the bees become gradually more and more inactive with the advent of the first frosts, until at the approach of winter they fall into the semidormant condition in which they exist until the spring sunshine rouses them to renewed activity.

This, then, is a brief résumé of a season's happenings in a colony of bees which are left to their own devices. Such devices, however, while well enough, no-doubt, from the bees' point of view, would be very detrimental, many of them, to profitable bee-keeping, so I will later endeavour to show how these wonderful little insects may be led into such paths as will benefit their owner without loss to themselves.

Old-fashioned bee-keeping consisted almost

entirely, of the let-alone system just described, terminating with the concomitant horrors of the brimstone pit and the garnering of a mess of indescribable components, by courtesy called honey. This honey was extracted from the brood combs by means of a press or the aid of fire.

The honey of to-day bears no resemblance, I am thankful to say, to the honey of our grandsires, while our modern systems are alike beneficial to bee-keeper and to bee.

CHAPTER II

STOCKING AN APIARY

THERE are several ways in which an apiary may be effectively stocked, and the advisability of adopting any one method is much a question based upon the time of year in which the work is to be carried out. Bees may be acquired in the form of (*a*) established colonies in frame hives, (*b*) as colonies in skeps, (*c*) as swarms, (*d*) as driven bees. In the case of established colonies, whether in frame hives or skeps, they may be bought at any time except during the winter months, say from October to March, when, as no proper examination of the bees can be made, it is not advisable to buy. The best time for purchasing these colonies is at the end of March or during April, when a warm day may be chosen and a satisfactory examination made. A novice is strictly cautioned against purchasing any stock on their own combs unless he has expert advice as to their freedom from disease, or a written guarantee from the vendor stating that the bees are perfectly healthy.

Good theoretical knowledge is of no use in

diagnosing disease. Foul-brood, for instance, that arch-pest of bee-keepers, could not be detected by one novice in a hundred in an incipient stage. I cannot lay too much stress on this point, for as one who has been through the fire more than once, I speak feelingly when I say that the acquisition of diseased bees by a beginner will in all probability, effectually quell all his aspirations in the direction of apiculture, and many a good man has thus been lost to the craft.

Colonies secured in April, if in fair condition, give ample time for getting the bees up to a point when they can take full advantage of the honey flow. Cases are quite common where the first season's honey has paid for the bees. A fair price for a colony of bees in a good hive is from 25/- to 30/- at this time of the year. For this money the buyer should receive a colony which covers from six to ten frames thickly, with brood on at least four frames, denoting the presence of a fertile queen. They should be absolutely healthy and the combs should be good. When speaking of good combs, I mean that they should be quite flat and free from drone comb. Healthiness should be a *sine qua non*, and they should not be pollen-clogged nor should they be too old. Colour indicates age, and in a light brown state, as distinct from dark brown or black, they will do. There should be a fair amount of brood in all stages and some stores, but if the other essentials are there the question of food supply may be waived and the bees fed with sugar syrup.

Skeps of bees should certainly not be bought by beginners on their own responsibility, and in fact it would be best if they ignored them altogether. Disease cannot be detected at all in a skep, as the combs are fixed and cannot be examined without cutting them out. An expert, if he wanted bees, would take the risk, cut out the combs and examine them, afterwards transferring them to a frame hive if healthy; if otherwise, destroying the combs and treating the bees as an artificial swarm or driven lot. This work, however, is quite beyond the beginner. If he buy skeps at all he must be prepared to take the risk, and this is not advised. If by any means he comes into possession of any, the safest way of dealing with them is to place them on a stand and allow them to swarm, hiving the swarms into frame hives. As skeps are only acquired nowadays as adjuncts to frame hives, this method is to be preferred even in the case of healthy skeps to the usual method of dealing with them, which consists in transferring the bees and combs as mentioned before. The state of the combs matters little in this description of hive. They should not be too black, but that is the only thing that need be insisted on. The usual price for skeps is from 12/- to 15/- each, but the latter is a top price.

Swarms are usually recommended for beginners,

and there is much to be said for such a recommendation, as swarms cannot transmit disease, and if they are obtained early they will often give a handsome surplus in their first year. Indeed, like good stocks, they will often recoup their owner for his outlay. They should be obtained as early in May or June as possible, and preference should be given to those weighing about six pounds. Swarms are usually sold by the pound, and it is much the best way of buying them. The price is usually 3/- per pound in May and 2/6 in June. It will be noted that swarms when received will weigh a little less than the weight charged for, especially if they have been sent from a distance. This is owing to the fact that when bees swarm their honey sacs are full, but when received by the buyer this honey has been partially consumed, and there is a corresponding loss in weight.

Swarms should be hived in a clean hive, and with new quilts and frames. There will then be no danger of disease of any kind; and the persistent way in which this method of starting is advocated is well borne out in practice.

"Driven" bees are utilized for conversion into stocks. To the uninitiated I may explain that "driven" bees are the bees taken from cottagers' skeps during August and September; bees which are saved from a cruel death in the sulphury pit, which unhappily still exists and has many votaries, especially in remote country districts. These bees may be purchased at about 1/- to 1/6 per pound, and it takes a six-pound lot, which sometimes means the contents of three skeps, to make a good colony for a frame hive. These bees are placed in a hive containing about six frames of foundation, or preferably of empty combs, and fed up rapidly with good syrup, when usually they will turn out a good colony in the spring. This method of founding colonies is not advocated unless the bees can be had for nothing or for a small consideration. This is only possible when skeps are personally driven, when the rule is that the driver takes the bees for a trifling acknowledgment and the skeppist takes the honey. This done, stocks may be made up cheaply, but when the bees have to be bought, along with comb foundation and sufficient sugar to feed them up for wintering, the total cost comes perilously near that of a prime swarm. Now a May swarm, nine times out of ten, will give a surplus, but "driven" bees cannot possibly give any return before the following season. Thus a season's working is lost; wintering risks have to be faced with bees which, be they ever so good, are rarely up to the stamina of a swarm for facing its rigours. Driven bees do not transmit disease.

To summarize, it may be said that, putting aside the proviso regarding gratuitous " driven " bees, the choice should lie between stocks and swarms, with stocks for preference if properly bought, but for untutored hands swarms, and nothing but swarms.

Apiaries should be situated if possible in a sheltered position. A small glen or valley suits them very well, and the hives should be screened from prevailing winds. A south-east or southerly aspect suits the bees best, and the hives should be placed on bricks, or cinders may be put down and a firm base made by these means. Weeds and long grass should be rigorously kept down, and if the hives are on grass land a scythe may be used with good effect in keeping down the grass. Where there is any choice a careful selection should be made as regards locality. Most localities are fair honey districts, but there are some which are twice as good as others, both as regards the quantity and the quality of the honey crop. Careful attention must therefore be given to this. What is wanted is a good clover or sainfoin district, which means wide breadths of these valuable honey producing plants, along with plenty of fruit blossom for early brood production. Other good honey districts have such local crops as heather, mustard, or beans. Now all of these are valuable, and the apiarist who is situated in their midst has most of his hardest work done for him.

CHAPTER III

HIVES AND APPLIANCES

PROBABLY the earliest hive of which we in this country have any record is the old straw skep, and even now this hive is in extensive use, especially in remote country districts. As a home for bees it answers its purpose admirably, but as a means to up-to-date honey production it is utterly useless, mainly on account of its fixed combs, which make it impossible for any manipulations to be carried out. Now the proper manipulation of the bees and combs is absolutely necessary if the best results are to be obtained, and such work can only be carried out by keeping the bees in modern hives.

It must be clearly pointed out, however, that many people have modern hives and get no better results than they would have done with skeps, in fact they would have been pounds in pocket had they adhered to that primitive method of hive architecture. That, of course, is their own fault. The simple fact that the bees are in frame hives will not of necessity make the bees produce more honey, in the case of an ignorant or negligent owner, for with such an owner they would do quite as well, or perhaps better, if he had them in a hollow tree.

The frame hive in its modern form is, roughly speaking, divided into two classes, and nearly all other types are merely modifications. First we have the double-walled hive on the celebrated " W.B.C." principle. This hive consists of a floor-board with sunk entrance and detachable splayed legs, brood-box, eke and loose outer cases, covered with a span roof. All the sections are loose, and there is a free air space between the outer and inner walls, which tends towards an even temperature in the brood-nest. This hive may be used for the production of either comb or extracted honey, and if the former a special section rack, known as the " W.B.C." rack, is used. In this rack the sections are worked in wide frames, which keep them clean, and also admit of their interchangeability in the event of a poor honey flow. In such a poor flow the centre sections will be completed while the outer rows are untouched, and by bringing the outer rows to the centre full racks may be obtained. It may be said that both comb and extracted honey may be worked for at the same time.

Our second type of hive is the well-known single-walled hive, and probably there are more of this kind in use than any other. Its name is rather a misnomer, as, strictly speaking, it is single-walled on two sides only; the other sides have inner walls placed for carrying the frame-ends, and on these two sides there is a dead air space. The supers for extracting purposes are precisely similar to the body-box in construction, but shallower, and there is a lift for the quilts, the roof being either sloping or span. The floor-board has a sunk entrance at times, while at others the entrance is taken out of the bottom edge of the body-box. When sections are worked on this type of hive extra lifts are needed to accommodate the section racks. These are generally of the form known as the "T" super, and contain twenty-one sections each.

Both single- and double-walled hives as now made are variable as regards outside measurements and in minor details, but in one respect they are standardized, and that is in respect to the size of the frames. The number of frames in a hive is ten usually, and it may be nine or eleven, but they will be of the outside measurements of the British standard frame. This frame is 14 inches by $8\frac{1}{2}$ inches outside measurement, with a 17-inch top bar. This bar should be $\frac{3}{8}$ thick by $\frac{7}{8}$ wide, the side bars $\frac{1}{4}$ inch thick, and the bottom bar $\frac{1}{8}$. These thicknesses are deviated from by various makers, but the outer measurements are inviolable. For supering purposes what is known as a shallow frame is used. This is identically the same as the standard frame except as regards depth; it is $5\frac{1}{2}$ inches deep only.

With regard to working qualities both these types of hives are admirable, and there is practically no difference in the results to be obtained. Scientifically the "W.B.C." is the better hive, and it is rather more pleasant to work with on account of its lightness and the detachability of all its sections. Against this, however, it must be pointed out that it is a difficult hive to secure for travelling, and a more expensive hive either to purchase or to make than the other variety. Where a few hives only are being kept for a hobby this hive should be purchased, but for a bee farmer the so-called single-walled type is recommended.

When commencing bee-keeping there are a certain number of indispensable appliances which must be obtained. This list, however, need neither be lengthy nor costly, but the articles obtained should be of the best possible quality, and of the simplest construction compatible with efficiency. Appliances should never be bought second-hand, except perhaps in the case of extractors and ripeners, which may be picked up in that way. All other goods should be bought new, and from a good firm.

A veil and smoker are two of the most necessary items, and it should be seen that the veil is large enough. Many veils are narrow and of poor material, an abomination to the wearer, and an inadequate protection against the attacks which it is their duty to ward off. The veil should always be black in colour, or it is not easy to see through, but white net may with advantage be used for the part which covers the neck, as being cooler during the heat of the summer.

The smoker should preferably be of the "Bingham" pattern, and of a good size and quality. The best articles are made of sheet steel, and these are to be preferred, for a poor smoker may be worn out in a season's working. Gloves should not be worn. Their use makes the bees more irritable than they would otherwise be. This is mainly on account of the clumsy way in which their owner handles the frames; clumsy he perforce must be with gloves on, for no skilful bee-work can be done except with the naked hands.

Other articles which will be needed are feeders (one for each colony), excluder zinc (one sheet for each hive), an extractor if extracted honey is to be worked for, sections and section racks if comb honey is desired. In the case of extracted honey a supply of supers and shallow frames will be needed, as well as spare standard frames for the brood-boxes.

One or two queen-cages are also necessary, in fact indispensable at times. As will be seen, a certain number of the articles in this list may be eliminated if the bee-keeper decides to work for honey in one form or òther only, and it may be said at once that it will be found in nine cases out of ten that extracted honey will be the best form in which to turn out the produce. The explanation of this will be given later.

With regard to this list it should be noted that

the feeders had better be of the "Universal" or graduated type, while the extractor should be of a size suitable to the number of colonies it is proposed to keep. A small extractor, costing perhaps 15/-, would be quite large enough for an apiary of half a dozen hives, but for a large apiary a large machine such as the "Cowan" should be acquired. This fine extractor takes four combs at once, and honey can be thrown out by the cwt. In buying queen-cages the pipe-cover variety will be found to be the most generally useful. They cost 6d. each, and if two are obtained it will be ample for a small apiary.

In buying frames, whether standard or shallow, the kind is largely a matter of individual choice. They are all alike in size, but differ somewhat in construction, which is not a vital point at all. If the buyer has not any particular preference, he may be recommended to get the ordinary dove-tailed frame with a saw-cut in the top bar, which, in the writer's case, fulfils all requirements very satisfactorily.

The other items mentioned will be described and commented upon in the chapters devoted to their use, as in that way the reader will be better in a position to understand their workings.

SPRING STIMULATION

With Bottle-and-stage Feeder

To face page 32

FINDING THE QUEEN

An Expert using the Hive Cover as a Seat during manipulations

CHAPTER IV

THE PREPARATION OF FRAMES AND SECTIONS

BEFORE new frames and sections can be used in the hives it is necessary that they should undergo a certain preparation, which consists in filling, or partially filling them with sheets of comb foundation. Comb foundation is pure beeswax rolled out into sheets of varying thickness and embossed with the imprint of worker cells, or more rarely, with the form of drone cells. This is done by passing the sheets through heavy rollers, which embosses and at the same time imparts an added toughness to them which is very useful in assisting the combs to withstand the high temperature of the hive without breaking down. The invention of this material has probably done more to revolutionize bee-keeping than any one thing, with the possible exception of the honey extractor. By its aid combs may be obtained perfectly flat, built entirely of worker cells, and with a very great saving to the energy of the bees, which saving is that of the bee-keeper also.

Before this material was brought out it was impossible to obtain, consistently, combs of even fair

c   33

quality. They, were crooked, and generally contained a large percentage of drone comb, which made them practically useless as a means of honey production. With the use of these wax sheets the bees find half of their work already done for them : the septum or mid-rib of their cells is in position, and all that remains to be done is to draw out the walls. Needless to say, comb building progresses at a rapid rate.

Comb foundation should be used with the greatest liberality, as liberality—not extravagance —in this direction is most profitable. The more work the bees can be saved in comb building the greater the gain, for note that bees consume from thirteen to twenty pounds of honey in secreting the wax for one pound of comb. Many, bee-keepers who are supposed to know their trade even now make a regular practice of placing starters in all their frames and sections. Starters are narrow strips of wax, about half-inch deep, placed along the top-bars, and their use, as their name indicates, is to give the bees a start. It does that and nothing more, and the resulting combs in the majority of cases give strict indications as to their origin. Full sheets of foundation should always be used, and not starters, or even half sheets. The sheets are made in several thicknesses for different purposes. There is medium and thick brood foundation for the brood or standard frames, medium and thin brood for the shallow or surplus frames, and thin super for the sections. The sheets are sold in sizes to fit the various frames and sections.

Before frames can be filled with foundation it is very necessary that they should be wired to give the best results. They may be used without wiring, but in that case there is a great danger of the combs falling out of the frames during manipulation, and of their being thrown out by the force of the extractor.

Wiring consists of passing two or three strands of No. 20 tinned wire across the frames, which wires are embedded in the wax sheets, holding them in position, and ultimately becoming enveloped in the combs. Three wires are sufficient for the standard frame, and two for the shallow, placed at equal distances apart, and running across the frames from one side bar to the other. The simplest way of wiring frames is to bore fine holes in the side bars—three or two as the case may be —and secure one end of the wire to a fine tack placed just beside the first hole and on the outside. Then thread the wire to and fro across the frame, tighten up and finish off on another tack placed outside the last hole. It should be noted that in the case of dovetailed frames the joints must be nailed, as the dovetails alone are insufficient to carry the weight of a heavy comb filled with honey.

The foundation should now be placed in position in the top bar, and this may be a little difficult where saw-cuts are the means used for holding the sheet. The best way is to open the cut by means of two small wedges placed at either end outside, then insert the sheet, which on removal of the wedges will be securely held. Note that the sheet should hang in the frame quite squarely, but just clear of the side bars, and a quarter inch short of the bottom bar. If it touches the side bars it will buckle, while if it reaches to the bottom there will be no room for "stretch," which exists to a small extent in even the best foundation. The wires may now be embedded, and for this purpose a board must be obtained of a size to fit the inside of a frame, and in thickness a shade under half the thickness of the frame to be fitted up. In the case of a frame having a top bar seven-eighths wide, a board about three-eighths thick will be right, and strips should be nailed across the back so that it may go no further than this depth into the frame. This board is now covered with a sheet of damp brown paper to prevent the wax adhering to it, and the frame with its sheet of foundation is fitted on to the board, with the wires uppermost. The best tool for the actual work is the "Woiblett" spur embedder, which consists of a toothed and grooved wheel, which is heated by a spirit-lamp and run along the wires, which are effectually secured. This little tool only costs a shilling, and is well worth the money. Care should be taken in using it to see that it is not too hot and that too much pressure is not used. Neglect of these precautions will result in the wax sheet being cut completely, through. Only just sufficient heat and pressure should be used to effect the object, which is the sinking of the wires well into the wax. This work may be done with an awl with a V groove filed in the point and heated in the fire, but the above-named tool is vastly superior and should be obtained.

In fixing foundation in sections the method to be adopted depends on the kind of section which is being used. Sections are made to take foundation in various ways. There are split-top sections, split-top and side grooves, split all round, and solid sections with neither split nor groove. The two sections most used, however, are the split-top and the solid section, and these shall be taken first. Section boxes are made in one piece, with V grooves at the joints and dovetails to lock the whole together. When they are bought they require folding, and before doing this a little warm water must be poured along the V grooves to impart elasticity, for they are liable to break if very dry. In folding split-top sections only one half of the lid should be placed in position before inserting the sheet of foundation, which is then secured by closing the other portion of the lid. Foundation for sections should be cut very, accurately. It must not touch the sides, and, as in frames, it should hang a quarter inch clear from the bottom. Being so exceedingly thin it will buckle at a touch, and buckled sheets of founda-

tion mean bulged and unsaleable sections when
completed. In using solid sections there is more
need than ever for accuracy in cutting the wax
sheets, for they are secured to the section on three
sides by molten wax. They should be cut with
a die of a size to fit exactly the inside of the
section, less a quarter inch at the bottom.   In
fixing the sheets a block must be made exactly,
the same as the one described for the frames,
and one-sixteenth less than half the width of the
section, and as the section will probably be two
inches in width, the block will be fifteen-six-
teenths. A little beeswax must then be melted
in a cup placed in a pan of boiling water, and
the section placed upon the block with the foundation in position. Now take a little wax in a
spoon, pour it along one edge of the wax sheet,
and by tilting the section the molten wax will
flow completely round and firmly fix the sheet.
This may seem difficult, like many other things,
but as a matter of fact it is very easy, and little
practice is needed to become proficient, while the
bits of appliances required will last a lifetime.

With the other types of sections the ways and
means of fixing the foundation will be fairly obvious when their construction is seen, and a knowledge gained of the previous methods. The main
thing is cutting the foundation accurately and
seeing that it is not nipped at any point, and
that means that the section-boxes must be folded
up squarely. After the boxes are folded they

should be put back into the racks which should be ready for their reception, and wedged up to keep all square, This is most necessary, as otherwise a lot of labour and foundation will probably be wasted. Empty section-boxes cannot be kept true unless they are properly squared up in the racks. After the bees have filled them with honey the case is different, as all the joints are then immovably fixed.

Finally, I would impress upon the novice the need for the utmost care and accuracy in this task of preparing frames and sections for the use of the bees. If this work is not well and efficiently performed, the work which follows afterwards will be equally bad. Combs will be twisted, or fall from their frames, while sections will be out of truth when filled, and eventually it will not be possible to glaze them. The blocks and dies must be accurately made and carefully used. If thin super foundation is "nipped" in the slightest degree, it will "buckle," or bulge in the centre, and a bad section is the result.

A frame is now being sold with a half-inch top-bar, and it has much to commend it. The ordinary bar, especially if it be split, is much too prone to sag under the weight of a heavy comb. This increases the space over the top-bars, with an additional amount of brace-comb, and other minor evils.

CHAPTER V

FEEDING PRINCIPLES

IN modern bee-keeping it is absolutely necessary that a certain amount of sugar feeding be done if the greatest possible amount of profit is to be derived from the bees. The duration-of this feeding and the amount of food supplied depend to a great extent on the method of bee-keeping practised by the owner of the bees, and also to a considerable degree on the district in which the apiary is situated.

Broadly speaking, feeding is practised for three purposes, which shall now be described.

First we have autumn feeding, the purpose of which is to supply the bees with a sufficiency of food to enable them to winter safely. Now in some districts little if any autumn feeding is necessary, on account of there being flows of nectar from certain late summer and autumn flowering crops peculiar to the districts. These crops enable the bees to gather a sufficiency of food for their needs, and as the supers will have been removed from the hives at the end of July, it is stored where it is required—in the brood-nest. With good bee-keeping, during the honey season proper little if any honey can be stored in the brood-nest owing to the management. The apiarist sees to it that the brood-combs are a solid mass of brood, and takes care there is a queen present who will keep them so. What honey is gathered therefore goes into the supers, which are taken off at the end of the regular honey flow, usually about the end of July. After that time what the bees gather they are allowed to keep. If there is a late flow to enable them to fill their depleted exchequer, well and good; if there is not, then they must be fed.

Again, to go to the other extreme, bad bee-keepers frequently have to do but little autumn feeding, and that is owing to the utter badness of their methods. These bee-keepers have many of their hives with comparatively worthless queens in possession, quite incapable of utilizing more than half of the brood-chamber for breeding purposes. Consequently the most of the early honey gathered is stored in the brood-chamber instead of in the supers, for be it noted that bees will never store in the supers while there is space in the brood-nest.

Spring feeding is food supplied for the purpose of inducing the bees to breed faster than they otherwise would, and is most important and even necessary in the case of apiaries situated in early fruit-growing districts, where bees must be very strong at an unusually early date if any surplus is to be secured. The amount of food given depends to a great extent on the quantity of stores with which the bees went into winter quarters, for it is neither necessary nor desirable to give fresh food until most of last year's stores are consumed, and if the bees are heavily fed in autumn it often happens that no fresh food is required.

The theory of spring feeding is this. Bees if left alone and with a sufficiency of stores will progress at their normal rate, which means that the hive will be at its best about the middle of June, when swarming will be in full swing. Now for the majority of districts, where the main honey flow is not due until then, true stimulative feeding is not required. All that is necessary is to see that there is always plenty of food in the hive, especially if bad weather intervenes in May and June. The districts that require stimulative feeding, however, have their main honey flow considerably earlier than the above date, so early that if left alone the bees would be much too weak to be effective, so the owners of bees in such districts must stimulate them.

The third kind of feeding is feeding that must be done at times, irrespective of weather or season, if the bees are to be kept alive. Of course in many cases it is negligence or bad management that calls for this kind of feeding, but in other cases it is not so. There have been, and there will no doubt be again, summers in which it was necessary to feed the bees all through on account of the utter wretchedness of the weather, which quite precluded any honey gathering. It must be noted that during the breeding season the amount of food consumed by a strong stock is enormous, and a few days of rain may bring it to the verge of starvation; in fact, inattention has killed thousands of stocks before now in bad years. The bee-keeper must make sure that there is always a supply of food in the hive if through adverse climatic conditions the bees are unable to gather a daily supply. It is quite easy for bees to be starved to death with thousands of acres of clover in bloom around them. Therefore when there is no food in the hive, feed, and feed until the bees can bring some in.

Now regarding the method of carrying out feeding we will take autumn feeding first again. When the supers are taken off in July, if there is no other late source from which a honey flow may be expected, it is a good plan to feed gently about a quart of syrup weekly to each stock, until the beginning of September. This will induce the queen to continue breeding later, and ensure the colony going into winter quarters with a strong force of young bees, which are a most valuable asset in wintering safely. The food should be given them through one hole only of a graduated bottle feeder. The feeder consists of a bottle

with a metal top perforated with nine small holes.
This is inverted on a metal plate contained in a
wooden stage. The metal plate is furnished with
a slot, and by turning the bottle round feed may
be given from any number of holes from one to
nine, or it may be withheld entirely. At the beginning of September this slow feeding should
be stopped, and food should be given as rapidly,
as the bees will take it, until the hives contain
thirty pounds each of food. This may be esti-
mated by examining the combs, noting that about
$4\frac{1}{4}$ inch square of sealed comb equals one pound.
For this fast feeding what is known as a rapid
feeder is often used, a box-like receptacle holding about a half-gallon of syrup, but bottle feeders
will do if the bees are allowed to feed from all
the holes. The recipe for syrup for autumn feeding is as follows: 10 lb. of best cane sugar,
5 pints of water, 1 tablespoonful of vinegar, and
a pinch of salt. Boil for a few minutes.

Spring feeding or spring stimulation is practised
entirely with the graduated feeder if continual
feeding is required. It should be commenced
at about six weeks before the expected honey,
flow, that being the time that an average stock
takes to get into condition. Food should be given
very slowly, from one or at most two holes of the
feeder, but it must be continuous. During the
whole of this six weeks the bees must be handling
food without ceasing, but they must not be storing
it in any quantity. While honey or syrup is com-

ing in, or is being manipulated by the bees, even in driblets, the queen will continue laying, apparently under the impression that there is a honey flow; but should the supply cease, even for twenty-four hours, then the queen ceases also, and the hive is the poorer by two to three thousand young bees. The supply must be regulated to the daily needs of the bees, and it must be given so slowly that it takes them the whole of their time to obtain it. Should it be given too quickly, and too fast for their consumption, they will store it in the combs, crowding out the queen, who will thus be restricted in her breeding.

When there is a heavy store of last year's food in the hive this should be utilized before more is given. The best way of doing this is to bruise a few inches of capping thrice a week. This can be done with a knife, when the bees will clear out the cells. When all the stores have vanished continue the feeding if necessary with the bottle. The recipe for spring syrup is: 10 lb. of best cane sugar, 7 pints of water, 1 tablespoonful of vinegar, and a pinch of salt. Boil for a few minutes. This recipe is used for food at all times during the spring and summer, but syrup must not be given between the months of September and April. If bees need food then it must be given to them in the form of soft candy, which may be obtained from any dealer in apiarian supplies.

It is not necessary, but it is very desirable, that all food given should be medicated with napthol-beta solution, as a deterrent to the attacks of foul-brood. This drug may also be obtained from appliance dealers in one-ounce packets, along with directions for the use of it.

Where Isle of Wight disease is prevalent, substitute "Bacterol" for napthol beta as a medicating agent, and on no account omit adding it to syrup. Few liberties could be taken in the treatment of foul-brood—none can be taken when this latest pest is in the vicinity.

Careful attention should be paid to the quality of all sugar syrup fed to bees, and also to the season of the year when it is given. Neglect of these details may set up dysentery in the colonies. This trouble is often caused by the feeding of low-quality sugar, and also by feeding it so late in the autumn that the bees are unable to seal it down. Fermentation then takes place, and is accentuated if syrup of thin consistency is used. The symptoms of dysentery consist of a great weakness of the bees, and a quick decrease in their numbers, combined with much soiling of the hives with excreta, both internally and externally. The treatment consists of shaking the bees into a clean hive, with fresh combs of good stores, or, failing these, candy should be used. Pack the bees up warmly, closing the hive up with a division board to the number of frames covered, and give ample ventilation. By such means the lowered vitality of the colony will be raised to the normal. Undue disturbance of the bees in winter also tends to set up this malady.

CHAPTER VI

THE PRODUCTION OF HONEY

THE main thing with nine-tenths of the people who keep bees is the production of honey and the pecuniary benefits to be derived from the sale of that valuable article of food. Now keeping bees and producing honey in large quantities are two widely different things, as many people have before now found out. Many people, in fact anybody, can keep bees, but it is not everyone who can become a bee-keeper. A keeper of bees and a bee-keeper are by no means synonymous terms.

A bee-keeper must be in sympathy with his charges: he must understand their peculiarities and be thoroughly alive to their every need, both immediate and prospective. The mere fact of having bees in the garden is not to be taken as an augury of honey in the cupboard, for in some seasons the production of honey calls for skill on the part of the bee-keeper of no mean order. I could take as an instance the season just past, when probably not more than one-third of the people who keep bees secured surplus honey worth the name.

To secure honey in quantity it is necessary that the stocks should be at their strongest just at the time when the main honey flow occurs in their particular locality—just at the time, neither before nor after. This is only to be secured by having good queens at the heads of the colonies, and by careful attention in the way of management as described in the previous chapter. At the time when the honey flow is about to commence the brood-chamber of all hives should be a solid mass of brood, and if it is not in that condition it must be made so. This may be done by uniting the stocks as described elsewhere. It is a common failing with the majority of bee-keepers to estimate the strength of their apiary by the number of hives containing bees. This is entirely wrong, for often half of the hives contain bees that are but remnants of colonies. Such hives are of no use at all for honey production, and left to themselves they might secure sufficient honey for their own consumption, but that would be all. If, however, they are united under one roof, making up a colony packed with bees and brood, good results will be secured if the season be favourable. That is the great secret of honey production— hives packed with bees and brood at the right time. Two colonies of bees under one roof will secure far more honey than if they were separate, and therefore as honey is what is required, unite,

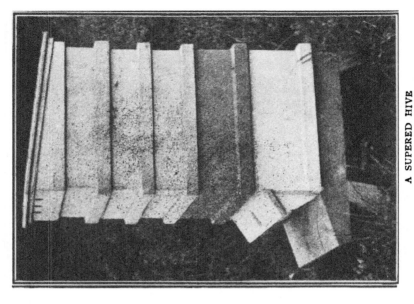

FOUL-BROOD

A portion of a badly-diseased Comb, showing various
aspects of the malady

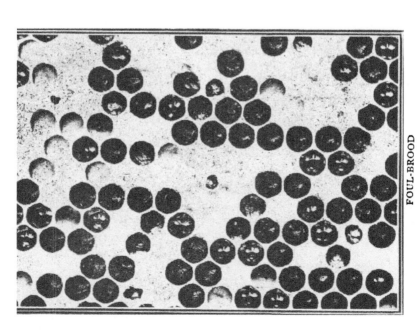

A SUPERED HIVE

Carrying Four Supers, all of which are occupied

To face page 49

and then if necessary divide the colonies again later for increase. Unite three or four lots if necessary to make a strong stock.

The bee-keeper should make up his mind as to what form he wishes his surplus honey to take, for on that will depend the whole of his supering arrangements. If he wishes to produce comb honey he will require section racks and sections, which must be fitted up in readiness. If he should prefer combs for extracting he must equip himself with the requisite number of supers fitted up with shallow frames.

Each hive should be furnished with two supers, and these will usually be sufficient for all requirements.

In the case of sections it should be carefully noted that they should always be used with separators. These are thin slips of metal or wood, which are placed between each row of sections. Their object is to prevent the sections being built out of shape and of varying thickness. Without the use of separators it would be impossible to obtain even approximately perfect sections.

The super is placed on the hives at the first indication of the honey flow, and this may be easily known by taking a glance at the brood-nest. If the cells along the top bars are being extended, which will be known by the snowy whiteness of the new comb, it is known that the bees need further storage room, and it should be given at once. In placing a super in position little skill

D ·

is needed, but it should be seen to that the tops of the brood-combs are scraped clear of projecting brace-combs and a sheet of excluder zinc laid over them, the slots of which should run at right angles to the frames. This excluder zinc is simply a sheet of slotted zinc which will admit of the passage of the workers but not of the queen. The object is to prevent the queen using the super for breeding, to the detriment of the shallow frames kept solely for honey producing purposes. This zinc may be omitted in using section racks, but it should always be used under shallow frames. The reason of this is that the queen has a great objection to passing through the narrow entrances of section boxes, but will freely pass up into a shallow frame super. Having the zinc in position, place the super directly down upon it, and cover all up as warmly as possible. When the first super is two-thirds full of honey the second one may be given, placing it under the first. By the time the third super is needed, if one is needed, the top one will be ready to come off.

Always place the last super given next the brood-chamber, that is, underneath all the others, except towards the close of the season. If, however, there are signs that the honey flow is failing, although the bees require more room, the last super may be placed at the top, when the bees will only use it if necessary.

The management when comb honey is worked for differs in no vital particular from the above, but it may be pointed out that it is of no use attempting to secure good sections unless the stocks are very strong and the district at least fairly good. It is impossible to produce good section honey in poor districts, and in such districts attempts are only a waste of time. It calls for much management to keep down the number of incomplete sections. The apiarist must be very, careful to give no more room than is absolutely, required and compel the bees to finish their work as far as possible as they go on.

In removing surplus honey from the hives the super clearer should be used. This is a flat board, the size of the hive top, with a " Porter " bee-escape in the centre. This, escape allows the bees to leave the super, but effectually prevents their re-entering it. This is an invaluable appliance, and should always be used when honey is to be removed. Evening is the best time to place it on a hive. A little smoke should be blown into the super, which may be then removed and placed upon the clearer, which should be in readiness by the side of the hive. Then replace the super on the hive, with the escape still underneath it, and cover up with the roof. The bees will all have left the super in twenty-four hours, and it may then be removed.

As regards the advantages of working for either comb or extracted honey, it must be confessed that most bee-keepers go in for the latter. It has many advantages which section honey does not possess. In the first place it is not every district that will produce sections, and then again when produced they must be consumed at once, or they deteriorate.

With the changed economical conditions and the great rise in food values, it is certainly most advisable to work for extracted honey. There is no doubt that as a food honey will be once again a prominent factor, but it is very unlikely that the price will again fall so low as in the past. Section honey is indisputably a luxury, and it is as an essential article of food that we would have honey looked upon. In this case it is not in accordance with the spirit of the times to produce comb honey of a perishable nature when we can produce a far higher quantity of extracted honey, which keeps indefinitely, and is obtained much more cheaply by the producer. Food values will without doubt remain high for many years, and bee-keepers should do their best to augment as far as may be the nation's food supply.

It should be widely known that honey taken from diseased colonies is quite wholesome, and may be used for household purposes in the ordinary way. Germs of bee diseases are absolutely harmless to human beings ; it is only necessary to keep the honey secure from the visits of marauding bees.

CHAPTER VII

NATURAL SWARMS

TOWARDS the end of the month of May swarms may be expected from strong colonies. As a rule there will not be many swarms during this month, by far the greater number coming off in June or during the first week of July. Still it is best to be prepared in time, and by this date spare hives should be in readiness for use if necessary.

There are various reasons which explain the swarming of bees, first and foremost being that it is their natural mode of increase. In a great measure this accounts for the fact that the numerous devices intended to eliminate swarming have never proved entirely successful. There are, however, other conditions which conduce to the bees seeking another home, conditions which may be swept away by the apiarist if he has no use for increase.

Chief of these is want of room in the hive for the teeming population, the exposure of the hive to intense heat, and the failure to give adequate ventilation. A colony is also more liable to throw off a swarm if it be headed by an aged queen. These incitements to swarming may be obviated by giving ample room for the requirements of the colony, the deposing of old queens, and provision for a sufficiency of ventilation and shade during hot weather.

If a hive swarms, however good the management a certain amount of honey is lost, so that the greatest amount of surplus can only be secured by the checking of this propensity so far as is possible. Much may be done to this end on the lines laid down, but steps must be taken in time. If no attention is given until the bees have decided to swarm, which is known by the fact of their building queen-cells, attempts at prevention, if not quite useless, are very nearly so.

It is most difficult to check swarming when queen-cells have been once started. The swarm usually issues at the capping of the first queen-cell, and is accompanied by the old queen. The bees will only issue if the weather be fine and warm, and the time usually selected is between 10 a.m. and 3 p.m. Should there come a period of adverse weather the bees may not leave the hive, but, tearing down the queen-cells, either await a more favourable opportunity or abandon the idea for the season.

When a swarm leaves the hive the bees fly forth in a thick stream, and after circling round for a few minutes in the air alight in a dense cluster, generally on a neighbouring tree or bush. As soon as they have become quiet they should be hived at once. If they are left alone, after a varying length of time they will again take wing. If this happens the swarm will probably be lost, as the bees may travel many miles and at a high rate of speed.

The operation of hiving is really very simple, and consists of merely placing an empty skep or box beneath the bees and dislodging them into it by a vigorous shake. Then gently place the receptacle on the ground, crown upwards, and raise it an inch or so by placing a small stone under the bottom edge. Soon the flying bees will have entered, when a cloth may be drawn over the mouth of the skep and the whole at once removed to the position to be permanently occupied.

It may be said that apart from the presence of queen-cells in the hive there are no certain signs that a colony is about to swarm. The bees are often listless and apathetic just previous to swarming, but this is not always the case, nor is it an invariable indication of their intentions.

It may be that the swarm will have clustered in a rather difficult position, as in the bottom of a hedge, or on the limbs or trunk of some tall tree. In such cases the ingenuity of the bee-keeper will be called into play to effect the capture. If the bees cannot be shaken into the skep they must be driven into it by means of a feather dipped in a solution of carbolic acid and water, or by the aid of a smoker. In nearly every case it will be necessary to adopt one of the two methods given, and when carbolic acid is used

it should be a 25 per cent solution of Calvert's
No. 5.

The transference of the swarm to the frame
hive should be left until evening. The frame
hive must be prepared with the full number of
frames filled with whole sheets of foundation, and
a division board. Cover the frames with quilts,
consisting of a sheet of strong calico at the bottom
with a three-inch circular hole for the feeder,
and three good felt wrappers above. Block up
the front of the hive about an inch from the
floor-board, and make a sloping platform up to
the entrance. This platform may consist of a flat
hive roof covered with a tablecloth, and on it
the bees should be thrown, as close to the entrance
as possible. Jerk them gently from the skep, and
when all have entered the hive-front may be
lowered. Owing to the heated state of the bees
it is best to remove all the quilts except the
bottom one for twenty-four hours, after which
time they may be replaced. Give food, however,
in the form of syrup, and continue to do so until
all the combs have been built out and honey is
available.

When replacing the quilts on the second day
contract the hive with the division board to the
number of frames occupied by the bees. Add
additional frames to the cluster when required,
until all have been drawn out. Then if honey is
abundant a super may be put on over a sheet
of excluder zinc.

One of the best methods of utilizing a swarm is to hive it on the parent stand, moving the old stock to one side with the entrance at right angles to that of the swarm. The next day turn the entrance of the parent colony, a little more towards its old position, and continue doing so each day, until at the end of a week the two entrances are side by side. Now remove the old stock to a new position, which will add a considerable number of flying bees to the strength of the swarm, and also discourage the throwing off of after-swarms on the part of the old stock.

Casts are second swarms thrown off by old stocks, generally on the ninth day after the first or prime swarm. They are headed by an unfertile or virgin queen. Usually it is best to return these swarms, after cutting out all the remaining queen-cells. Other after-swarms may be thrown off if not prevented, in addition to the casts. These small swarms should always be avoided by the removal of the cells, as they greatly weaken the parent stock.

Casts and after-swarms will issue in any kind of weather, and frequently travel far afield.

When swarms are required for sale due attention should be paid to their packing, especially, in the matter of ventilation. They travel best either in the original hiving skep, or in a specially constructed swarm-box.

If they are despatched in the skep the mouth should be tied over with a single thickness of cheese-cloth, and the skep placed upon its crown and secured in an open box.

Special swarm-boxes are constructed with large openings in the sides and top, which openings are covered with perforated zinc, providing a plentiful supply of air. In all cases label conspicuously, "Live Bees, With Great Care," and despatch to the customer at once. When sending by rail they should always be sent by passenger train.

When swarm-boxes are used it simplifies matters if the bees are hived directly into them. This can easily be accomplished by darkening all the ventilation openings round the box by means of brown paper or cardboard tacked on the outside. The box then can be used in the same way as a skep, but be careful to uncover the ventilation openings before sending the swarm away.

With Isle of Wight disease so rampant it is very inadvisable to send bees from one district to another, for there is no doubt disease is largely spread in this way. No one can definitely say that their apparently clean bees may not be spore carriers, and for the time being, at any rate, it would be well for all swarms to be sold at home. One effect of the ravages of this disease is a sharp rise in the price of bees, in the shape of either swarms or stocks. These cannot now be obtained at the prices quoted in other chapters. The prices now are in a sense abnormal, and while it may be a considerable time before we get back to the old prices—we may not get quite back to them at all—there will without doubt be a considerable fall as the supply of bees again reaches the normal.

CHAPTER VIII

MARKETABLE PRODUCTS

THERE are, in modern bee-keeping, various ways in which the stock may be turned to profitable account. To utilize to the full these numerous opportunities of making money should be the aim of every progressive bee-keeper. It is only, however, in isolated instances that it can be said that the most is being made of the apiary.

For selling purposes we have in an apiary saleable stock and products. Taking the products first, these may be subdivided into two classes. In the one class we have articles produced more or less in every apiary, and which consist of comb and extracted honey and wax. In the second class we have articles of food, etc., manufactured by the bee-keeper to supply requirements in his trade, such articles containing a certain amount of either honey or wax.

Among these articles special mention may be made of mead, honey vinegar, furniture cream, lip salve, honey soap, and a large variety of cakes and confectionery.

As regards the stock which may be sold, this consists of swarms, stocks, nuclei, and queen-bees. With reference to the manufactured articles there is really nothing to prevent any bee-keeper making and creating a market for their sale, providing that he finds that it pays him best to do so. He must, however, make one branch of the pursuit particularly his own, the other branches being looked upon as merely subsidiaries. As to which particular branch he favours, this is largely a matter of individual preference, tempered by the quality of the district and the class of trade which is to be catered for.

In some good districts, where fine light honey and good sections can be produced in quantity, it would probably pay the best to take that line. In another district, that produces an abundance of inferior honey, the apiarist should go in for the selling of stocks, swarms, and queens. An early honey district, too, is very good where swarms are looked upon as a chief source of income. In these districts the swarms come off very early, and make a price which is considerably in advance of those of a later date. In making his plans the bee-keeper must never forget that he cannot produce and sell both bees and honey in quantity unless he has an exceptionally large apiary. It must be one or the other in most cases.

If swarms are allowed to come off, if the best colonies are sold, or if queen rearing is largely indulged in, very little honey will be secured— probably none unless heavy sugar feeding is adopted to replace any that is taken away. .

Where honey is produced it will usually be found advisable to go in for the extracted article, on account of its superior keeping qualities, although where there is a good demand for sections there will be little difference from the point of view of profit.

Good extracted honey in bulk will readily sell at sevenpence per pound, while first-rate sections will make about ninepence each in dozen lots. Against this must be put the fact that bees will produce about thirty per cent more of extracted than of comb honey, so that the prices are fairly even.

Section honey must be sold quickly or it granulates in the comb, and becomes unsaleable on this account. This form of honey should not be extensively produced until it is known that there is a ready sale for it. There is such a large demand for good honey that little more need be said at this point. The bee-keeper will find that his greatest difficulty as a rule is to cope with his orders.

Beeswax is a very valuable article, and is produced to some extent in all apiaries. For this also there is practically an unlimited demand by dentists, chemists, and similar professions. The very high melting-point of this wax makes its use imperative in some of the mechanical processes of various trades, and every scrap of comb should be religiously preserved for melting down and converting into cash. There is not a very large quantity of British wax thrown upon the market, owing to the fact that much of it is converted into foundation by the bee-keeper for his own use.

Wax in bulk sells for about 1/8 per pound, or 2/- if it be melted into one-ounce cakes for retailing over the counter.

The utilization of honey and wax in the manufacture of articles for domestic use or for food is becoming an increasingly large industry, and I know at least one extensive bee-keeper who has built up a large trade. This line is very useful where, owing to the district, a crop of dark honey is secured. Dark honey makes but a low price on the market, but answers admirably for the making up of confectionery; better, in fact, than the light honey of more delicate flavour.

This class of trade has of course to be created and built up by canvassing shopkeepers and by advertisement, but when once the trade has been secured it is very remunerative. Recipes for various articles may easily be obtained, but I give two excellent ones for mead and honey vinegar.

In selling swarms the motto should be to get them as early as possible, and for this reason it is well not to give the bees too much room. Eight frames will be ample, and every incitement should be given them to swarm in May. Swarms should be sold by weight, which is the fairest way for both buyer and seller. If it be a May swarm, charge 3/- per pound; if the month of June or early July, 2/6 is the usual price.

Colonies of bees may be sold at any time of the year between March and October, although they are much more valuable in spring than in autumn. A fair price for a good stock in spring is 25/- to 30/-, but in the latter part of the year they are not worth more than £1. Both colonies and swarms are very saleable, in fact I never knew a season of late when the supply was equal to the demand. In selling stocks be sure that they are free from disease, and if they are sent off by rail great care must be exercised in packing them. Bees on combs travel badly.

The packing of a frame hive will be found described in the chapter dealing with heather honey. If stocks are sold off, the supply must be kept up by raising new ones each season. This may be done by forming nuclei early in May, giving each of them a queen-cell, and building up these small lots into full colonies.

The sale of queens is becoming a very important item in apiculture, but it is a branch in which a reputation has to be made, and this often takes several seasons of hard work. The most expensive part of queen-rearing is in the fertilization of the virgin queens, and where many are raised nearly the whole of the stocks in the apiary have to be broken up into nuclei for this purpose. Queens fetch a good price in early spring and sometimes in late autumn, and the breeder must lay his plans to meet this demand.

Much care and attention is needed in queen-raising, and none except really superior stock should be sold. A fair price for ordinary fertile queens is 5/- each from March to May, 4/- in June, and 3/6 at other times.

Recipe for Honey Vinegar. Take one and a half pounds of honey to a gallon of water, a crust of bread, and a tablespoonful of brewer's yeast. Place these in an earthenware barrel, and stand in a warm place. After fermentation has ceased cover the bung with a piece of linen to exclude insects, and allow the liquor to stand until ripe.

Recipe for Mead. Take six gallons of water and one gallon of honey. Boil until it is reduced to four gallons. Add half-ounce of ginger, quarter-ounce of cinnamon, half-dram each of cloves and peppercorns in a bag. Boil for a few minutes longer, and then let it stand until fermentation has ceased. Add yeast if necessary to assist fermentation. Barrel, and bottle at the end of twelve months.

EXAMINING SHALLOW COMBS

Showing Honey partially sealed

To face page 64

CHAPTER IX

CORRECT MANIPULATION

THE great majority of the people who keep bees have much to learn as regards the proper handling of their stock. This is usually due to sheer ignorance, owing to the lack of a little tuition or the want of a reliable text-book or periodical. It is rarely through any unwillingness to learn, as the modern methods are in every way cleaner, quicker, and far more pleasant than the antiquated forms of apiculture met with in many secluded districts. Occasionally one comes across some hoary-headed unbeliever who persists in keeping to the traditions created by some remote ancestor, but the breed is fast dying out, and bids fair soon to become extinct.

The great secret of manipulating bees is to be gentle but firm, and the more the former quality is cultivated the less need will there be of the latter. Bees are very nervous insects, and object exceedingly to anything in the way of roughness. In opening a frame-hive always stand at the back of it if possible, or at the side, but never in front of the entrance. A veil must be worn, but no gloves, and some form of subjugator will be required. This usually takes the form of a smoker, or of a cloth sprinkled with a solution of two parts of water to one part of Calvert's No. 5 carbolic acid. First remove the quilts until only the bottom one remains. Gently raise the edge of this and blow a little smoke over the frames, afterwards replacing the quilt. The effect of this smoke is to frighten the bees, causing them to fill their honey-sacs from the open cells. In this condition they are much less liable to sting.

If there are no open cells containing honey, a condition of things which often exists in early spring or late autumn, a little sugar syrup must be poured over each of the seams of bees. It will take the bees a couple of minutes or so to fill their sacs, after which time the quilt may be removed and a little more smoke blown over the combs. Any examination that may be required can now be made. In taking out the frames, first remove the division-board at the side, if there is one, and then take out the first frame.

In handling frames there is one correct method, and one only. Draw the frame a little sideways to clear it from the next comb, and then gently raise it by the lugs. Examine the side nearest to you, and then, lowering one hand until the top bar is perpendicular, swing the frame round like the leaf of a book and bring the hands level again. The effect of this will be that you are now look-ing at the other side, with the bottom bar upper-most. To bring the frame back to its proper position reverse the movements. Other and quicker methods will occur to the novice, but there is no other way, in which the comb is absolutely, safe. In the manner described it is impossible for the comb to fall out of the frame, and this is a very, real danger, especially in summer when the wax is soft and the combs heavy with honey. The result of a heavy comb leaving the frame and falling into the hive is not readily imagined, but it is always a severe lesson.

When replacing the frame in the hive be careful that it does not drop hard on to the runners, and place it against the near side of the hive, after which examine the other frames in strict rotation. Two thorough examinations annually, are quite sufficient for ordinary purposes, one in spring and the other in autumn. Careful notes should be made at these times of the state of the colony as regards strength, the age of the queen, and the condition of the combs. Especially should close scrutiny be made for signs of disease.

A hive should not be kept open longer than is necessary, and in case the bees get restive they, should be subdued with a little smoke. Novices should note that it is not always necessary to hunt the queen up, providing that her presence is assured, and of this worker brood and eggs are a sufficiently sure sign. In replacing the quilts, first smoke the bees from the top bars, and then lay the wrappings on one by one. If a carbolic cloth be used instead of a smoker, it should be drawn over the frames as the bottom quilt is peeled off. In a minute or so remove it, only using it further when the bees show signs of insubordination.

When the spring examination is made, and it should be made on a warm day towards the end of April, a clean hive must be given to each colony, afterwards scraping out the old hives and washing with strong soda and water applied very hot.

In examining straw skeps, the combs being fixed, their removal is out of the question. All that can be done is to blow a little smoke in at the entrance and then invert the skep. The combs may be prised apart with the fingers, when the presence of worker brood will indicate that the skep contains a fertile queen. Nothing further however, can be done as regards examination, and often the weight is no very correct guide to the amount of stores contained, as pollen-clogged combs may account for much of it.

In manipulating bees it will be seen that their temper varies a great deal. They are much quieter at some times of the year than they are at others, and they are easier to handle on warm days than they are on cool ones. It is unwise to examine them late in the evening, and the best time of all is in the middle of a warm day, when --honey is coming in. It will be found also that they are especially irritable at the close of a honey flow, say at the end of July and August, when great care is needed in handling them.

It is of course to be expected that occasionally the apiarist will be stung, although his chief troubles will commence if his bees sting the neighbours. It is the fear of stings that deters thousands from taking up this most fascinating pursuit, but really a sting is a mere nothing to most bee-keepers. It is all pure imagination chiefly. The flesh swells a little with some people on first being stung, but even this discomfort usually disappears after a short time, until except for a little momentary pain there is no ill effect whatever. In many cases I often receive stings without noticing them. There are exceptional cases where people are constitutionally unable to bear the effect of stings, and where a single sting even causes most serious symptoms, but such instances are very rare.

With regard to treatment, most apiarists after removing the sting ignore it altogether, but if it be thought advisable, a little ammonia, washing blue, or soda may be rubbed on the affected part. Brisk rubbing should be strictly avoided, or a painful swelling may be caused, owing to the diffusion of the poison. The sting must always be removed at once by a scratching motion of the finger-nail.

Beginners in bee-keeping should try and gain a fair idea of the condition of a colony from outward indications. This knowledge can only be acquired by close observation and experience, but it will save a considerable amount of manipulation, and be it noted that manipulation is not good for bees, and the less they have of it the better.

The strength of a colony may be gauged by the number of the flying bees ; the presence of a queen by the way in which they carry in pollen ; want of food will be known when larvæ are cast forth ; robbing will be easily discovered ; the presence of disease indicated, and even a desire to swarm occasionally foretold. The alighting board of a bee hive is an open book to those who can understand its language, and many things of which I have no space to tell will be found printed there.

Study carefully your hive entrances, and watch the movements of the bees. You will thus save yourself the trouble, and the bees the annoyance of many unnecessary disturbances of the colony. It is here that you will often see the first dread signs of Isle of Wight disease, and be able to take immediate steps for effectively dealing with it. The different colours of the pollens will tell you what are the sources of supply, or the lethargic demeanour of the bees betray the fact that the honey flow has ceased, and that there is no work to do. The experienced bee-master rarely pulls his colonies to pieces, but he never ceases his watch over the entrances of the hives. Never open up a colony if you can possibly discover what you require from outside indications.

CHAPTER X

MID-SEASON WORK

AFTER his supers are on the bee-keeper must redouble his vigilance, or he will be liable to suffer loss in various directions. Particular care must be exercised in seeing that the bees have plenty of room for storage purposes, and space must always be given in advance of their requirements. The exact amount of space required is a matter which can only be learned by experience, but it may be said that an abundance may be given to strong colonies at the commencement of the season. Towards the end of the honey flow much more care must be used, or a lot of unfinished work and unripe honey will be the result. When the honey flow is on the wane, which is easily seen by a peep into the supers, give extra room very sparingly. Empty supers of combs should not then be given, except to very strong stocks, but full combs of capped honey may be taken from the centre of supers and their places taken by empty combs.

If too much room is given at this time the bees will scatter the honey about, using a small portion of each of many combs, and filling none, a form of storing which makes much extra work for their owner. This question of getting all work finished off applies in a much greater degree to sections. Unfinished sections are unsaleable. At the best they can only be extracted, and it is poor economy to extract sections. To ensure their being finished off, the finished sections should be taken from all the racks as soon as the supply, of nectar is seen to be failing, and full racks made up of the unfinished ones that are left. These racks must then be placed upon the very strongest stocks and wrapped up warmly. By these means the bees will be induced to complete them.

Swarming is usually the chief trouble to many bee-keepers at this time, and every means possible should be used to avoid this trouble if a large amount of surplus is required. The best method for checking it and dealing with the swarms has been pointed out elsewhere, but there are a few other points well worth mentioning. One of these is that if sections are worked for, the number of swarms will be much above the average. Bees intensely dislike these little boxes, with their tiny, cubicles and general lack of space. It is not a natural way in which to compel them to build comb, and consequently the great majority of such stocks throw off swarms. These swarms must be treated strictly on the lines indicated, as regards hiving them on the old stands and giving them the supers.

The common way of dealing with a swarm is to hive it on a new site, and more often than not the supers are left on the old stock. This is not a bit of use, and the adoption of this method will lead to a loss of nearly the whole of the honey crop.

A few spare queens may be easily reared when a swarm comes off, by the simple plan of breaking up the old stock into three nuclei; each nucleus consisting of three frames with the adhering bees, and one or two good queen-cells, of which there will usually be an abundance. The division is best made about six days after the stock has swarmed. In making it, allow one nucleus to remain on the stand of the old colony, and make the two others on new locations. The one remaining on the old spot may be left rather weaker than the others when the division takes place, as it will be strengthened somewhat by a number of flying bees which will return to it from the others. In a few days time the queens will have hatched, and on becoming fertilized may be usefully employed for re-queening purposes, or for sale. The nuclei may then be joined up again to make a full colony.

At the end of the honey flow, unless every care is taken, there will be much trouble from robbing bees, and when bees commence to rob in earnest the apiary rapidly becomes demoralized. It very soon becomes the despair of its owner and a terror to the neighbours. Fighting takes

place on a large scale and many bees are killed.
The strong stocks rob out the weak ones, and
when these are finished off fight fiercely among
themselves in their endeavours to rob each other.
As a rule one strong stock cannot rob out another
which is equally strong, or even moderately strong,
if the weaker one has a little assistance from the
apiarist.

The great thing is to check robbing at the very
commencement, to nip it in the bud before the
business has time to get fairly under way. If it
becomes a serious case it is very difficult to put
it down, and it certainly means the removal of
nuclei and weak colonies to another apiary for
the time being.

Robbing is nearly always caused by careless-
ness on the part of the apiarist himself. He
throws bits of comb about on the grass, keeps
the hives open too long, or leaves supers of honey
thoughtlessly exposed. Honey or syrup is spilled
about the apiary, and the deed is done. The
trouble commences at the close of the honey flow,
and if it is fairly started it will often go on in-
termittently until the frosts put an end to the
predatory warfare.

It should never really get a start, and certainly
ought never to get out of hand if due care is
exercised. Honey must be removed from the
hives in the evening, and examinations of colonies
made at that time also whenever possible. The
bees will then have ceased flying. No sweets of

any kind must be left about exposed to attack, and · all hive entrances should be contracted. Strong colonies may have a two-inch entrance, but weak colonies and nuclei must have one bee space only.

If a colony be attacked sprinkle the alighting board with diluted carbolic acid, and throw a bunch of loose, wet grass over the entrance. A sheet of glass may also be propped before the entrance. This will help to baffle the robbers in their attempts to force a way in. If these methods fail, the attacked stock must be removed for the time being.

It may be remarked here that when stocks are moved at a time of the year when the bees are flying freely, they must be taken a distance of at least two miles, or many flying bees will return to the old site and be lost. If it is necessary for any purpose to remove them a short distance at such times, the hives must be moved at the rate of two or three feet on each fine day. This, however, would not be practicable if for any purpose it was necessary to move a stock a distance of a mile. In such a case they would have to be taken to a spot three miles distant, kept there for a week or so, and then placed in the desired position. In the winter, when the bees have been closely confined to their hives for a period, they may be moved either long or short distances without loss.

The cleaning-up of combs wet from the extractor will be a part of the work which has to be done at this period, and this is a fruitful cause of robbing. The supers of wet combs must be placed on the hives at nightfall, and care taken to see that no bees can effect an entrance from the outside. The combs may be allowed to remain on the hives for a week, at the expiration of which they may be removed and stored away.

It is recommended that this "cleaning-up" of wet combs should be entrusted to one or two colonies, as by this means there is not so much danger of distributing disease germs. The usual method is to give the combs back to the colony from which they were taken, but it must not be forgotten that such combs have usually followed others through the wet cages of an extractor. Where the extracting is done by taking one colony at a time, and cleaning the extractor for each, I would allow each colony to clean its own combs, but not otherwise.

We are speaking now of an apiary in which there is no known disease. It is only to such apiaries that the above remarks apply. Where any mild cases of foul-brood exist—a few infected cells, say, here and there—the honey from such stocks must be treated quite as a thing apart. In such cases extract the honey from the healthy colonies, and finish with the others, finally disinfecting the extractor.

CHAPTER XI

THE time has undoubtedly arrived when no intelligent bee-keeper can honestly resist the introduction of an Act of Parliament dealing with bee diseases. The advent of Isle of Wight disease (*Microsporidiosis*), with its terrific ravages, makes such an Act a vital condition of successful apiculture. Formerly the apiarist considered his troubles to be infinite when his apiary contracted foul-brood, but this old brood malady pales beside the fierce and deadly pestilence which is now devastating our apiaries.

Briefly, the chief foes of the apiarist are Isle of Wight disease, and foul-brood, and he should be thoroughly conversant with the symptoms, aspects and treatment of these maladies. He has other troubles, dysentery, bee paralysis, and May pest, while at times toads, tits, and a few other birds will prey upon his bees, and wasps will steal his honey. Mention is made of dysentery in an earlier chapter, while the depredations of the toads and other small fry are usually infinitesimal, and may be easily checked. In the case of bee paralysis, and May pest, there are grave suspicions for

77

assuming that these are but mild forms of Isle of Wight disease.

Isle of Wight disease is caused by the presence of a small parasite in the chyle-stomach, and intestines of the bee. This parasite (*Nosema apis*) after passing through various stages, forms spores, and it is by means of spores that the disease is spread. It is a complaint which is very difficult to diagnose at times, as the symptoms vary considerably. In mild cases it may take the form of 'spring dwindling,' or that of a bad case of dysentery, but in its more serious phases it is no longer open to misconstruction. Several things may happen even then, however. Frequently the bees are found dead in a heap on the floor-board of the hive, when the spring examination is made, and at other times the bees vanish entirely, and are never seen again. The very commonest form is the spectacle of great numbers of bees crawling about on the ground, and ascending blades of grass and other objects, unable to fly, and exhibiting distortion in several forms. The abdomen may be distended, and appear to hang downward, wings are often projected from the thorax at unnatural angles, and one or more pairs of legs may be paralysed. Occasionally, the combs are soiled with excreta, but not always, although any fouling of the hive interior gives good ground for suspicion. The worker bees are the first usually to be attacked, and the queen is, as a rule, the last to fall.

It will thus be seen that there are many aspects, but the progress of the disease is very rapid as a rule, and any bee-keeper who is losing many bees, in conjunction with any of the symptoms mentioned, will be correct in assuming that the trouble is Isle of Wight disease.

As regards treatment, it is to be regretted that no certain cure has been found for infested stocks, and most of them perish. Something may be achieved in mild cases by feeding the bees with food medicated with Bacterol, and by spraying the bees with a solution of the same preparation. Serious cases should be dealt with in the most drastic manner. The stocks should be destroyed, and the bees, combs, and all interior hive fittings burnt, although any honey or wax may be used for household purposes if desired. The hives should be scorched out with a painter's lamp, washed out with a five per cent. solution of carbolic acid, and the outsides re-painted. The ground under and about the hives should be turned over, and sprinkled with lime, or carbolic solution, and a good supply of fresh water should be assured. Stagnant water is a fertile source of infection.

In combating the disease, absolute cleanliness is essential, and all means should be adopted to maintain the utmost vitality in the bees.

Foul-brood (*Bacillus Alvei*) differs from Isle of Wight disease in that it is a larval disease, but it is little less formidable on that account, and takes

a heavy toll of bee life. The bacillus, after certain
changes, forms spores, analogous to the seeds of
plants. These are easily disseminated, and of
great vitality, retaining their power of activity
after a period of years, and they are immune to any
chemical agent which would not injure the bees.
When attacked the larvæ turns to a pale yellow,
and later to a dark brown substance of glue-like
consistency, and with a very objectionable odour.
In the early stages the larvæ which is unsealed is
noticeably flabby and distorted, while the sealed
larvæ in bad cases shows sunken and dark-coloured
capping, often perforated by the bees with irregular holes, as if an attempt had been made at
removal.

Black brood is a very similar disease, and in bad
cases both of this and foul-brood, the same mea-
sures should be adopted as for Isle of Wight
disease. Milder cases may be treated by the
starvation method. This consists of shaking the
bees from their combs, and confining them in a
box, or skep, for forty-eight hours without food.
Give them plenty of air, and at the expiration of
the time hive them in a clean hive on sheets of
foundation, and slow-feed with syrup medicated
with napthol beta. If possible re-queen the
colony.

In dealing with these two disorders, it must
never be forgotten that both are highly infectious,
and the spores may be readily carried from colony
to colony by means of contaminated appliances.

HIVING A SWARM

ANCIENT AND MODERN HIVES

Here will be seen an old straw Skep in company with a modern "W.B.C." Hive

To face page 81

CHAPTER XII

INCREASING AND UNITING

IT frequently happens that it is desirable to divide colonies, either for increasing the stock or with a view to the prevention of swarming. The method usually adopted is the one known as artificial swarming, and it is capable of many variations. Artificial swarming is especially useful in cases where it is suspected that colonies may swarm, for then this may be done for them, and any possible loss of the natural swarm or trouble in securing it obviated. There are certain rules to be observed in this process of division. They are few in number, but they must be rigidly adhered to, or failure will attend the efforts of the apiarist. First, only strong colonies must be divided; secondly, the swarm must be made in the middle of a warm day, when the bees are flying freely; thirdly, it must not be done before drones are plentiful for the fertilization of the young queens.

To divide one colony into two, take the frame on which the queen is found and place it in a new hive, filling up the hive with frames of empty comb or sheets of comb foundation. A frame must also be placed in the first hive to replace the one taken out. Now remove the full colony to a new location, and on the vacant site place the hive containing the queen. The bees which are out gathering, along with others which will return from the removed hive, will make up the swarm, which should be well fed until established. The old stock will raise a new queen, but if a fertile queen can be given, or even a ripe queen-cell, valuable time will be saved.

Where one stock is made into two no honey is secured that season as a rule, and the next variation is recommended as giving moderate increase with a possibility of surplus honey as well.

This form of division consists in making three colonies out of two. Select two strong colonies, and from one of them take five good frames of brood. Place them in an empty hive and fill up both hives with empty combs or sheets of foundation. No bees must be taken with the brood, and the hive containing it must be placed on the stand of the other selected colony, moving the latter to a new position. Thus the bees are secured from one stock and the brood from the other. In this case it will be observed that it is the new colony which has to be given a queen, or if necessary allowed to raise one. If it be desirable the frames of brood may be taken from any number of colonies up to five, when their loss will not be felt.

We will now reverse this procedure, and turn to

uniting. This operation becomes an obligation at times, as in the case of weak stocks which are unable to winter, or with nuclei at the close of queen-rearing. It is also necessary with moderate colonies just before the honey flow, so that they may be placed in a position to take advantage of it.

When bees are united care must be taken to prevent fighting, and if precautions are not taken the weaker of the two parties will be killed. The best plan with weak colonies is to gradually move them towards each other until they stand side by side. Remove all empty combs until each colony occupies the same number of frames as nearly as possible. Take away the worst of the two queens, and cage the other upon a comb with a pipe-cover cage. Now dust the bees thoroughly with flour in the hive containing the queen, and space the combs wide apart to admit of the others being placed alternately with them. Then dust the queenless bees and place them in the spaces reserved for their reception. Take away the empty hive, cover up the united stock, and liberate the queen in twenty-four hours.

Syrup scented with peppermint may be used instead of flour if preferred. Never attempt to unite " driven " bees by running them into the entrance of an established colony, or they will be killed. If it is necessary to use these bees for strengthening purposes, hive them first upon combs, and then proceed upon the lines indicated.

With the plan given, any number of nuclei may be joined together, and if there is no choice in the queens it is not necessary to remove them. The bees will settle that to their own satisfaction without any trouble on your part. It is only advisable to cage the queen when all the others have been removed, to avoid any risk of her being damaged.

When disease exists in a locality it is wise to refrain from interchanging brood and bees from established stocks. No more colonies should be interfered with than is absolutely necessary. In making up a nuclei, for instance, if three are required, break a colony up entirely, but do not make one from each of three colonies. This forming of nuclei colonies early in the season is a very good method of increase, but care must be taken or they will suffer checks, and fail to build up sufficiently strong for wintering. They should be gently fed when there is no natural supply, and carefully protected from robbers.

Instead of increasing native stocks it will be advisable to add to the number of Dutch colonies for the present. These Dutch bees, while not immune from the attacks of Isle of Wight disease, have very strong constitutions, and are great disease resisters. Unfortunately, just now the demand for these bees exceeds the supply, but with a cessation of war conditions this would probably adjust itself.

CHAPTER XIII

QUEENS AND QUEEN-CELLS

THE importance of young queens in an apiary cannot be over-estimated. They are a necessity if the utmost profit is to be obtained from the pursuit, and yet no phase of bee-keeping has less attention paid to it in the majority of cases. Why the queen is so important is obvious, but what is not generally known is that queens, like most other things, deteriorate as they get older, until they reach a stage when they are no longer profitable. A queen is at her best in the second season, and she will probably do well in her third, but after then it is not advisable to retain her. Thus, then, theoretically we ought to renew our queens at the end of every second season or thereabouts, but personally I am no advocate of such an arbitrary system. While a queen continues to do well I should retain her to the third season, but I should have a younger queen ready to depose her if she failed before that time.

When a queen is past her prime the bees will depose her themselves, first building a series of cells called supersedure cells, and raising a young queen. The bees, however, may drive this work on to a time long past that defined by the apiarist as his limit, and therefore it is not wise to leave it to them.

When bees are found to be building supersedure cells, by all means let the good work go on and utilize the cells, as they produce the best of queens. Good cells may also be obtained when a colony swarms, but at other times when the apiarist wishes to rear queens he must compel his bees to raise a batch. This is done by making a colony queenless, and at the same time depriving it of all unsealed brood and eggs. Twenty-four hours afterwards give them a frame of eggs from the hive containing the best queen in the apiary, first cutting the comb away at the bottom up to the first row of eggs. This will give room for the queen-cells to hang. Feed the queenless colony gently, and in about ten days' time the nuclei may be formed.

Nuclei are small colonies of bees formed for the purpose of getting the young queens fertilized, and they should consist of three good frames of bees, two of the frames containing mature brood. The queen-raising colony will form three good nuclei, one of which, and the weakest of the three, should be left in the original hive. Nuclei may be placed either in full-sized hives or in small hives holding three frames only.

Give each of these small colonies at least one good queen-cell, enclosed in a cell-protector, and cover the whole up warmly. In a fortnight from that time, if all goes well, the young queens should have hatched, have become fertilized, and be laying. They can be used as may seem desirable, and a fresh batch of cells given to the nuclei.

It will be seen that although a queen raiser can breed from his best queen, he cannot select the drones with which the young virgins are to be mated. Much has been written about the fertilization of queens with selected drones, but in a country so thickly populated as our own the thing is practically impossible. Fertilization takes place high in the air, and even if the bee-keeper kept nothing but selected drones in his own apiary, a large proportion of his queens would probably be mated with those which belong to his neighbours.

When introducing new queens to a colony precautions must be taken, or they may be killed. The usual method is to cage the queen on a comb, placing the cage in the centre of the brood-nest and over a few open cells of honey. Feed the bees gently, and liberate the queen in from twenty-four to forty-eight hours. It will be readily seen on releasing her whether the bees are disposed to be friendly or otherwise towards her. If they commence to pull the queen about, or enclose her in a tight mass of bees, a form of attack termed "balling," she must be caged again until they are willing to accept her.

Another method of introducing is to enclose the queen in a cage, at one end of which is a quarter-inch hole filled with soft candy. The

queen is then released by the bees themselves
eating away the candy block.   In any case be sure
that there is no virgin or other queen in the
hive, or any attempt at introduction is bound
to end in failure.

For commercial queen raising it is usual to
make up the nuclei in small hives, taking frames
half the size of the standard, thus economizing
the bees required for the fertilizing stage.   These
small frames are clipped together and used as
ordinary standards during the early part of the
season, afterwards taking them apart when forming the nuclei.   System is the secret of profes-
sional queen raising.   Everything must as far as
possible work smoothly, and batches of cells must
be continually brought forward to replenish the
nuclei from which fertile queens have been sold.
This branch of apiculture is not for beginners,
and should be adopted very cautiously and gradually, extending operations as the demand increases
and experience ripens.

In the presence of disease, the hands should
be well washed with carbolic soap between the
examination of one stock and another, and the
appliances should be disinfected with a twenty-
five per cent. solution of carbolic acid.   If your
stocks are healthy it is a most unwise thing to
introduce fresh blood into the apiary at the
present time.   Many thriving apiaries have been
ruined by neglect of this precaution.

HONEY EXTRACTOR

Chain geared and fitted for four shallow Frames

UNCAPPING HONEY

The Capping falling away before the knife, owing to the sloping position of Comb

CHAPTER XIV.

THE PREPARATION OF HONEY

WHEN the honey has been removed from the hives, much still remains to be done before it is fit for the table or the market. In the first place, if it be extracted honey that has to be dealt with, it requires grading, extracting, ripening, and finally putting up in jars or other receptacles.

This work must be carried out in a room which is impervious to the attacks of bees or wasps, as it usually comes at a time when these insects are highly aggressive. Should they be able to effect an entrance the work will soon be brought abruptly to a standstill.

The honey should first be graded, as it is a mistake of the first order to extract the whole of the combs in one lot. The honey which is collected by the bees is of various colours and qualities, and it must not be mixed if the best is to be made of the product. If the combs be held up to the light it will readily be seen that they contain honey of both light and dark shades, and these shades should be separated and extracted apart. Light honey will fetch threepence or fourpence per pound more than dark honey, although for my own part I much prefer a good honey of a medium colour. However, the fact remains that light honey is synonymous with high prices, and in this respect there is no dark honey, that can compete with it save and except the incomparable heather honey of the Scottish moors.

This grading of honey must be done very carefully, for a very little dark honey will spoil the colour and bring down the price of a large quantity of light. The combs should be tested in a good light. After grading the extracting must be proceeded with, and for this purpose an extractor will be required, and also a few other necessary articles. The honey extractor consists of a metal cylinder, within which revolve two or more cages fixed to a vertical central shaft. The honey is extracted by uncapping and placing combs in the cages, which are then caused to rotate swiftly by means of a handle at the top or side, which handle in most cases is fitted with chain or cog gearing. The honey is thrown out by centrifugal force against the sides of the cylinder, and runs to the bottom, whence it is drawn off by means of a honey valve. All extractors work on this principle, and the machines run in different sizes, taking either two or four combs as a rule to each filling. As to which size is necessary depends on the amount of work which has to be performed and the price which the buyer wishes to pay, which may be anywhere from fifteen to fifty shillings.

Having the extractor, then, the other articles needed are two large sharp knives—the W. B. C. uncapping knife is best, but carving-knives will do well enough; a jug of hot water with which to heat them, or preferably a tin of water kept hot over a small oil stove; a tin for the cap-pings; and one or two cloths. Place the knives in the hot water, and then take a comb, which should be held in the manner depicted in the illustration. Now wipe the water from one of the knives, and remove the capping with one sweep-ing upward cut. The knife must be changed for the other side of the comb. As will be seen the comb is held at an angle so that as the cap-ping is cut it falls away from the comb and into the tin. After uncapping both sides place the comb in one of the extractor cages, after-wards uncapping another of about the same weight for the other cage. Now turn the handle sharply until all the honey is thrown out, after which the combs must be turned and the operation re-peated to extract the other side. Do not turn the handle too quickly, or the combs will be forced from their frames. Only sufficient motion is required to effect the purpose in view, so that the speed required will be very readily seen.

Continue the work until all the combs have been extracted, raising the machine as often as may be required to draw off the honey, and taking care not to mix the various qualities.

The honey should be drawn off into a honey-ripener. This is a tall, cylindrical vessel, shaped something like a milk churn, and is fitted with a strainer and lid. Its purpose is to ripen the honey. When honey is extracted there is a considerable quantity thrown out of the cells which has not been sealed over by the bees. This honey is unripe, and if it is bottled along with the ripe honey will cause fermentation.

The ripener is for the purpose of separating this thin honey, which should not be sold, but may be fed back to the bees. Having filled the ripener, then, place it in a warm corner and allow it to stand for a week. It will now be found that the thin honey has risen to the top, and as all ripeners are fitted with valves at the bottom the good honey may be drawn off, leaving the thin behind.

We now come to the putting up of honey for sale, and the methods employed are various. First let me say that cleanliness and neatness, combined with good taste, are the great secret of it all. Take whatever form of package you like, but your ultimate success in finding a market rests in the main on your observance of the qualities mentioned.

For the retail trade, or for shopkeepers, extracted honey is usually sold in pound and half-pound screw-cap jars. These jars are fitted with a cork wad, and if tastefully labelled they are particularly attractive. For the wholesale market it is customary to run the honey into fourteen or twenty-eight-pound tins, with a lever lid. Larger sizes than these should not be employed, as they are very heavy to handle, besides being inconvenient in cases where it is necessary to liquefy the honey after granulation. Square tins are the most convenient for packing where the produce has to be sent over railways.

Comb honey in sections also requires grading, and the combs should be sorted into first and second qualities. The first quality, consisting of the very best combs, fully capped over, built out to the wood all round, and of snowy whiteness, the seconds well filled but lacking in finish. After grading the wood must be scraped round with a sharp knife, to remove any marks of propolis, pollen, etc., and then for a finish it is best to glaze the sections. This is effected by attaching a square of glass with lace paper to each side of the section, using strong paste for the purpose. The glass should be four and three-sixteenths square for four and a quarter inch sections, and both glass and lace paper may be bought very cheaply.

It pays to glaze sections. In the first place they will make at least eighteenpence per dozen more than unglazed ones; then, again, the glass protects the combs from damp and the attacks of insects; finally, many shopkeepers will not buy them unless they are glazed. Comb honey is of a most fragile and delicate nature, and should have some protection if it is to be kept at all. In good condition it is a most beautiful and dainty article of food ; badly put up, sticky and messy, it is most uninviting and practically unsaleable.

Another way of putting up extracted honey is in the form of compressed paper, or fibre jars. These are after the form of the jars used for sending out cream, and are very cheap and serviceable. The paper is waterproofed, and the package is made in various sizes ; some which I received lately were made to hold two pounds of honey, and were fitted with a lid displaying the name and address of the producer. These paper jars are filled with the liquid honey, and the lid closed down, when they may be packed to travel by rail. Of course when the honey has granulated very little packing is needed, the paper being of a very tough and impervious nature.

This package is a very good one indeed for heather honey, and for any honey which granulates rapidly and solidly, as charlock blends. It is not quite so good as a jar for poor honeys of thin consistency, or for some of the fine clover honeys, many of which remain semi-liquid for years.

It is a good plan when selling honey wholesale to send screw-cap jars out in returnable boxes, and each box should contain a dozen jars, with separators between. Such boxes will stand for years, are a great aid to labour saving, and prevent losses from breakage.

CHAPTER XV

HEATHER HONEY

HEATHER honey is considered the finest honey produced by many experts, and it is certain that their opinion seems to be generally endorsed, judging from the large number of people who are prepared to pay the high price charged for the product. There has never been a season in recent years when the supply was equal to the demand, and the prices secured often range from 1/- to 1/6 per pound for the pressed article, and 1/6 to 2/- each for good sections.

This honey is of a very dark amber colour, aromatic, and exceedingly thick. In a good sample it much resembles jelly. The best qualities are secured from high altitudes, such as the Scottish moors, and parts of Yorkshire and Derbyshire. From lower levels the honey is of an inferior character. Heather honey is a most uncertain crop, and in most years there is a heavy shortage. The bloom comes at a time when the weather is often very unfavourable to the secretion of nectar, the nights being chilly, and there is frequently much rain.

As in working for clover the stocks must be

very, strong, and the body-box solid with brood when the bees are sent to the moors. If there is much room below, most of the crop will be stored there. Should this happen it is practically lost to the bee-keeper, as this honey cannot be extracted in the ordinary way. It is usual to work for sections, and the colonies should be strengthened until there are at least sufficient bees to fill one crate in addition to the brood chamber. Very warm packing is absolutely necessary, or the bees will not work in the supers.

It is a great advantage if the apiarist can get his sections drawn out during the clover flow, as this saves valuable time and is true economy. Bees are usually sent to the moors on a light spring cart or dray, in which they should be placed above a good thickness of straw.

When packing these strong colonies great care must be taken to ensure free ventilation, and the means employed are the same as when sending by rail. The frames should be secured from movement by screwing strips of wood over and across the lugs, and the body-box must be secured to the floor-board by screws also. Take out the entrance slides and close the entrance with a piece of wire gauze, and either gauze or cheese-cloth should be stretched over the frames after the removal of the quilts. This gives thorough ventilation.

Unless they are needed to accommodate the bees it is best to place the supers in position after

TRANSFERRING A SKEP

Skep placed above a Frame Hive to admit of the bees transferring themselves

To face page 96

arrival at the heather. In placing the gauze over the frames it is an improvement if it be mounted on a light frame made of half-inch wood, which will give a good bee-space above them.

If the apiarist makes a regular practice of going to the moors, he should have a three-inch hole cut in the bottom-board, and covered with perforated zinc. A cover may be made for this hole, which is only required at this particular time. The journey must be made at night, and the bees released early next morning.

As I have mentioned, heather honey cannot be extracted, but must be pressed out. For this purpose a honey press is needed. In these presses the combs are enclosed in cheese-cloth and crushed between metal plates. The honey is thus extracted and strained at the same time, ready for ripening and bottling. As this method involves the sacrifice of the combs most bee-keepers elect to work for sections in preference.

A bee-keeper is doubly fortunate when circumstances give him access to two such crops as clover and heather. He has not only a double chance of securing surplus honey, but even in years when the heather bloom is a comparative failure he usually secures sufficient honey to feed up the bees for the winter, and this alone is well worth the trouble and expense of the journey to the moors.

G

CHAPTER XVI

DRIVING BEES

ALL apiarists should be able to drive bees, and especially those who reside in country districts where skeps are much in use. Skeppists are often only too willing to allow their skeps to be driven at the end of the honey flow, and the acquisition of these bees supplies the modern bee-keeper with an opportunity of strengthening his weak stocks at a nominal cost. I do not advise the buying of driven bees, but where they can be procured for the trouble of driving, or for a small consideration, they are most useful.

There are two methods of driving, open and close, and it is with the former that we have to deal. The latter method is not largely used now, but I may say that it consists of inverting a full skep upon its crown and turning down upon it another empty skep of the same size. The junction is secured with a cloth, and by drumming upon the bottom skep the bees may be caused to ascend into the one placed to receive them. There is too much working in the dark with this method, however, which with open driving does not exist. Practically all that is required for open driving is a set of driving irons, and some skeps, linen bags, or boxes for the reception of the bees.

Gently smoke the skep to be driven, and turn it upon its crown, afterwards attaching an empty skep upon it as given below. The skeps are pinned together with the short iron, the others being used to support the sides. The point of junction, which forms the bridge by which the bees will reach the top skep, must be at the end of the central comb. This is a most important feature. The combs must run away from the operator, not across him from left to right, or there will be trouble. Place the skeps so that the strongest light comes from behind the operator, and then commence to steadily beat the sides of the bottom skep with the open hands. Soon the bees will begin to ascend, and in a few minutes the skep containing the combs will be clear, after which the irons may be removed and the bees shaken into a box or bag, leaving the empty skep ready for the next lot. If the bees instead of ascending show a disposition to boil over the sides of the skep, they must be checked with the smoker and induced to take the right direction.

These bees are useful either for the strengthening of weaklings or for the forming of new colonies. In the latter case about six pounds of bees will be needed, or roughly the contents of two or three skeps. If necessary two or three lots of bees may be placed in the same box at the time of driving, first removing any queens which are known to be old. It is best to hive these bees on built-out combs where possible, and feed them up rapidly. If this cannot be done, five or six sheets of foundation must be given to them, and they may winter on a contracted brood-nest of that extent.

Comb building at this season should be spared them when possible. It is to the advantage of the bee-keeper to obtain his bees as soon after the beginning of August as he can, and to this end he should be early afoot—or perhaps I ought to say awheel—among the outlying districts. Many old skeppists will probably be averse to having their bees taken up at this early date. Then it should be explained to them that except in odd districts, or in heather localities, the skeps get lighter nearly every week after the end of July, and this they may easily prove if they take the trouble to weigh them.

It is not a bad plan to go round in early spring and make arrangements with cottagers for the driving of their bees at the end of the season, and if a trifling amount per skep is offered there will usually be little difficulty in settling the matter.

In hiving the bees they should be treated exactly as swarms, and then fed as rapidly as they will take the syrup. A good set of driving irons may be made by taking some stout wire, about one-eighth thick, and bending it to the desired form.

CHAPTER XVII

SAFE WINTERING

WITH each recurring spring a large percentage of colonies fail to respond to the roll-call, and their disappointed owners cast vainly round for some explanation of what is to them an unintelligible mystery. The bees were all right in the autumn, strong, full colonies, and now they are dead. The same old thing repeats itself year after year, and will continue to repeat itself until bee-keepers more thoroughly understand what is required in order that their bees may winter safely.

Stocks ought to, and will winter safely, and I will guarantee that ninety-five per cent might be brought through. Personally I only admit of possible loss in one direction, and this is owing to the bees becoming queenless during the winter. There is no remedy for this. Queens, like everything else, die at times from natural causes, and if this occurs in winter the stock becomes a total loss.

All these lost colonies, however, of which I speak are starved to death in nineteen cases out of twenty. This may seem an astounding fact, but it is so, and when I state that I know of an apiary in which seventeen stocks out of twenty-four met with this fate last winter, nothing further need be said.

Bees can be wintered much safer than many other creatures on a farm if a few simple rules be thoroughly understood and acted upon. In the first place, they must have from twenty-five to thirty-five pounds of good food; secondly, they must be kept very warm in a good hive, from which wet, draughts, and mice must be rigorously excluded; finally, the colonies must be strong in bees and headed for preference by young queens, although this last is not an item of the first importance as regards wintering.

Preparations for wintering should begin at the end of August, when all colonies should be gone through and the food-supply carefully gauged, recollecting that a little over four square inches of comb contains about one pound of food. Make sure at the same time that there is a fertile queen present. When all the colonies have been examined proceed to give them any food that may be necessary to make up the requisite quantity for winter use. This should be given in what is known as a rapid feeder, a drawing of which will be found elsewhere.

When feeding is over, place a split ball of napthaline in each of the back corners of the floor-board, and across the frame top bars lay two strips of wood, about two inches apart. These strips should be three-eighths of an inch thick, and of a sufficient length to go right across the centre of the frames. Their use is to give the bees a passage over the frame tops, enabling them to travel from comb to comb in search of food without leaving the warm atmosphere which pre-vails at the top of the brood-chamber. This done, pack all down as warmly as possible with good thick quilts. Finally, secure the roofs from high winds, and if necessary place a strip of per-forated zinc across all entrances to exclude mice. Be sure that the hives are waterproof, as damp is fatal to bees.

If these instructions are thoroughly carried out, winter losses will become a thing of the past, and what was formerly a game of chance will be practically a certainty.

In connection with this chapter a word may be given as to the proper care of surplus standard and shallow combs during the winter months. These combs, which are most valuable stock, must be safely stored away, either in supers or in boxes specially made for the purpose. In any case they must be secured against the attacks of mice, and it is necessary that they be kept dry. The greatest enemy is the wax moth, the larvæ of which do much damage to disused combs. Their depreda-tions may be easily seen, and when they are present the combs should be well fumigated with burning sulphur. Prevention, however, is better than cure, and if the combs are kept in a receptacle which is moth-tight there is small fear of any harm coming to them.

A wintering system, which has much to commend it, is one known as the claustral detention system. In this method the hives are fitted with a special porch, which can be closed at will, and the bees confined to the hive during inclement weather in spring, or at any other time should the need arise. Ventilation is provided for by means of special tubes, and many other advantages are claimed for this hive entrance, which was introduced by the Abbé Gouttefangeas. It is asserted that bees have been confined for so long as five months in this way, and it is certainly very valuable in case of. moving bees to new locations, and in the making of artificial swarms, the prevention of robbing, and the eradication of bee diseases. In this latter case all that is necessary is to close the hive up entirely. Then no robbers can gain entrance, nor can the rightful inhabitants escape. There is no fetching and carrying of disease, and at the same time the medicinal treatment of the affected colony can go on.

CHAPTER XVIII

SELLING THE PRODUCE

THERE are a great many bee-keepers, mostly those who have small apiaries, to whom the selling of their produce at remunerative prices is an annual source of difficulty. These men, living as they do often in good but remote honey districts, can produce honey much more easily than they can dispose of it, while others having practically an unlimited demand are at their wits' end to supply their customers. It is quite time that this state of things was remedied, and some scheme launched on a co-operative basis whereby honey can be distributed evenly over the markets. The schemes which are now being inaugurated by the *Smallholder* would no doubt embrace something of the kind, as there are infinite possibilities in it, both for the selling of honey and any other kind of produce grown on small-holdings. As the thing stands at present, one man is producing large crops with no market for them, owing sometimes, I confess, to his own lack of business energy, while another man is running here, there, and everywhere in vain attempts to satisfy his numerous and ever increasing customers.

Without further digression, however, I would say that there is no crop which can be more readily sold than bee-produce, whether it be honey, or wax. Most bee-keepers fail owing to causes which are very obvious. They are deficient in push and energy, they reside in outlying districts and make no serious attempts to reach the towns, their goods are badly prepared for sale and offered in a messy and undesirable condition. These are the prime causes of failure, for I never knew a good bee-keeper yet who could not sell his honey; in fact, when a reputation for good honey has been made, people will readily fetch it, without causing any trouble to the apiarist.

First it is most essential that honey should be put up in faultless condition, and of a good level colour. Instructions for grading it have been given elsewhere, but the bee-keeper should try and make the grades approximately the same each season. Further, he will do well to establish a fair and regular price, to be adhered to without variation, either in good seasons or in bad. Create and establish a standard article at a fixed price. Put it up in the same way each year, until people can recognize your honey as far off as they can see the bottle. Never send out honey of inferior quality, or in a badly labelled, sticky jar.

If these precepts are adhered to, a market will be founded in an incredibly short time, and always remember that every satisfied customer usually brings others in his wake. Circularize the district and advertise in the local newspaper, at the same time making a small but tasteful display, in your window, if you be living by the side of the highway. If this does not clear your produce, recourse may be had to the nearest town and honey offered to the shopkeepers there, either by means of circulars, canvassing, or advertisement. Should this fail most shopkeepers will make a display on sale or return terms.

Try by all means to deal direct with the consumer, as then the highest possible profit is made.

With regard to the packages, there is nothing better than the usual screw-cap jars for extracted honey for the retail trade, and twenty-eight-pound square tins with lever lids for the wholesale market. All comb honey should be glazed preferably, or failing that it may be wrapped in grease-proof paper, and placed in cardboard cartons stamped with the name and address of the producer.

All honey should carry a distinctive label, containing the name and address of the seller, and to this should be added a notice stating that if the honey granulates it is merely a further sign of its purity, and that it may be readily re-liquefied by immersion in warm water. These hints if acted upon will do away with nearly all honey-selling troubles, which should never exist, seeing the large yearly influx of foreign honey.

CHAPTER XIX

RACES OF BEES

A QUESTION which is often asked by bee-keepers, especially just after they have left the novice stage, is, Which is the best variety of bee for me to keep? I invariably answer that a good strain of the common brown bee cannot be beaten.

Among honey-bees there are a number of varieties, but, except when kept experimentally, three only are commonly met with in this country. The ones usually seen are Italians, Carniolans, and our own brown bee. From time to time other races have been experimented with, but none of them have become popular.

The Italian bee is a brightly coloured insect, with an abdomen marked with distinct yellow bands. In the pure state it is exceptionally quiet, a prolific breeder, and an energetic worker. Many people here profess to have great faith in its good qualities, but candidly I have no great love for them personally. They have the good qualities I have spoken of, but they also have other bad ones, and one of these is that they are not of the slightest use for working on section honey, owing to the peculiar water-soaked appearance of their capping. If this was their only defect I could forgive them; but it is not, for while admitting the prolificacy of the queens, my pure stocks never attained the strength nor did they show results equal to my blacks. This I lay entirely to the delicacy of the adult bees, which caused their death-rate to be far higher than with other colonies. They do not seem to be sufficiently hardy for our fickle Northern climate, but I firmly believe in their great value for crossing with and improving the native race.

Carniolans are bees very similar in appearance to our own, and by many would be taken for such. They have, however, a more greyish appearance, and are much quieter in disposition than brown bees. Their queens are very prolific, and the workers are energetic and build splendid sections. Their worst defect is that they are inveterate swarmers, but, like Italians, they are very valuable for crossing purposes.

All these Eastern races have the swarming trait very strongly developed, and this feature causes considerable trouble to their owners, for if left unchecked some colonies will swarm while there is a pint of bees left in the hive.

Coming to our own brown bee, which is not a pure race at all, I do not consider that it can be beaten, provided that a good strain is secured. They are not so gentle as the foreign bees, I admit, but they are good workers, fairly prolific, and do excellent work on sections. A good strain is what beginners should endeavour to obtain, and when he has it let him yearly breed from the best queen in the apiary, when in a few seasons he will have bees which will beat any of the imported races. It will often be found that the best strains have a little Italian blood in their composition, which may make them of a rather fiery nature, but this is a detail if the balance sheet comes out right.

There is another side, however, from which the pure bee question can be viewed. Bees cannot be kept pure in this country, for it is too thickly populated, and there are too many bees. If pure stock is imported, owing to the peculiar method by which queens are fertilized, it rapidly becomes crossed, and in his endeavours to keep his bees pure the beginner often ignores the important feature of strain altogether. He finally finds himself with a lot of three-quarter bred Italians, of no particular merit as honey gatherers, but demons to sting.

Mind, I am not decrying these bees if they be used in their native land, as there is no doubt that in their proper element some strains are first-rate. What I do maintain is that in this country they are inferior to our own, and if, as sometimes happens, foreign bees of an inferior strain are secured, the comparison becomes more striking, so I adhere to my conviction that success is far more likely if a good strain of the native bee is secured.

CHAPTER XX

APPLIANCE MAKING FOR AMATEURS

OUR ideal bee-keeper is a handy man to whom very little comes amiss. It is an axiom that "bees do nothing invariably," and their many little tricks and vagaries would fill a book many times the size of this volume. To deal properly with the many curious circumstances which crop up calls for a certain amount of adaptability and ingenuity in the making, or improvising, of special appliances to suit the needs of the moment. Under

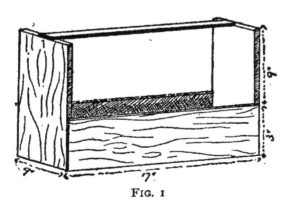

FIG. 1

these circumstances I am satisfied that most people who are able to keep their bees in order will find but little difficulty in making the appliances given in this chapter. This work, while affording pleasant amusement, will fill in profitably the long winter evenings, when little else could be done. The articles given by no means exhaust the list, for a great many other things could also be made with but little trouble.

A point I would emphasize is the necessity of accuracy in the measurements. A quarter of an inch more or less does not matter in many things, but in bee goods everything must be just right, or it will not be satisfactory. Then, again, it is recommended that the wood be bought ready, planed on both sides, which will bring it to an even thickness. If the amateur has to do a large amount of rough planing the keen edge will be taken off his enthusiasm before he gets to the really interesting work.

The first example (Fig. 1) is a most useful article, and is designed for the purpose of carrying the tools round from hive to hive, and also to act as a comb-rest. It is necessary at times to remove the first comb in a hive to give freedom in manipulation. In such cases the removed comb can be suspended on the stand, instead of placing it upon the ground, as is usual, to the danger of chilling the bees.

Very little need be said about the details, as the drawing makes it fairly clear. It should be made of $\frac{3}{4}$ inch stuff, and well painted, while the various parts may be either screwed or nailed together.

The super clearer (Fig. 2) should be carefully, made. The measurement, 18 inches square, is given on the drawing, but this may be modified by circumstances. The outside measurement of

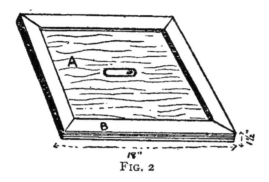

FIG. 2

a super clearer should be the same as that of the body-box of the hive on which it is to be used. The central portion (A), which extends right through to the outside, is formed from two ½ inch boards glued together. Round the edges, on both top and bottom sides, are placed the strips B, which are mitred at the corners and attached with screws. These strips are 1⅜ inches wide by ⅜ inch thick, and their purpose is to form bee-ways. The appliance is finished by fitting a " Porter " bee-escape into the centre. These escapes may be bought for sixpence each. A better way of making this article is to make a solid frame, 1⅜ by 1¼ inches, plough-grooved on the inner side to take the central portion of the clearer. This will prevent any "twisting."

The rapid feeder (Fig. 3) is on the Canadian

H

principle. In this the joints must be a very
correct fit. They must all be screwed and white-
leaded. Unless this is done the feeder will not
be tight, and the syrup will escape. This feeder

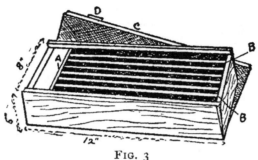

FIG. 3

may, be made of ½ inch wood for the outside, with
⅜ inch for the inner partitions. First make the
sides and ends, the former of which must be
grooved to take the piece A. Groove the end B
and the piece A to take the six slats, after which
fasten the outside together.

Now attach the bottom, cutting slots in it as
at A (Fig. 4). Place the screws very close

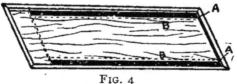

FIG. 4

together at the bottom, especially where the cen-
tral portion is secured to the two inner walls
BB (Fig. 4). The ends and the inner wall A
(Fig. 3) are all of the same level at the top,

finishing ½ inch below the level of the sides. The pieces BB extend right to the bottom, but the four slats in between finish ¼ inch from both top of A and the bottom. The walls BB finish ¾ inch below the sides of the feeder, and ¼ inch below the level of the piece A (see diagram, Fig. 5). The piece A must be cut away ⅛ inch deep in the centre of the bottom edge, as shown in the diagram (Fig. 6). The lid C (Fig. 3) is of ½ inch stuff, fitted with two ledges D to keep it from twisting. Round the bottom of the feeder, and on the outside, tack strips ½ inch wide by ¼ inch thick, to give a bee-way underneath. The

FIG. 5   FIG. 6

appliance is used by drawing the lid aside, exposing the reservoir at the end, into which the syrup is poured. From thence the food flows under the piece A and into the various divisions forming the central portion. The bees ascend through the openings AA (Fig. 4), and are prevented from drowning by the slots, which at the same time admit of a great number of bees drawing upon the food at one and the same time.

This is a very good feeder for autumn use, and it may be improved by lining the inner portion with tin. If well made, however, thickly screwed and leaded, a very serviceable article will have been secured. When finished melt a few ounces of paraffin wax and place it in the feeder, turning it about so that it may flow over all the joints. This will effectually fill any crevices there may be.

It is possible that some readers may possess straw skeps, and would like to know how to fix up a modern super for them. Years ago a bell-glass was considered the correct thing for a skep, but it is now possible to buy modern supers containing either shallow frames or a crate of sections. It is a very simple matter to make such a super, taking an ordinary super from a frame hive as a guide. The only difference is that a bottom must be nailed on, with a hole in the centre corresponding with the feed-hole in the top of the skep. Over this hole a small piece of queen-excluding zinc must be fastened. The sides of the super should be a little higher than the frames in order to carry the quilts—say 7 inches over all, instead of 6 inches as in an ordinary super, and a light roof is needed. The two inner walls, which carry the frame ends, should be removable, when comb honey may be worked for if desired. In placing the super in position, it will facilitate matters if a piece of felt is placed between the skep and the super, as it will rest more solidly, especially if a brick is placed upon the roof.

CHAPTER XXI

HIVE MAKING

THE hive (Fig. 7) is a good type of what is known as a single-walled hive. This hive is most efficient as regards its working, and it is also simple and economical to make. It is square as regards outside measurement, thus enabling it to be used with the frames either parallel to the entrance, or at right angles, as the owner wills.

The material used should be $\frac{3}{4}$ inch red deal, although other woods may be used. The hive floor board is designed for placing upon brick supports, but legs may be easily added if desired. First cut the pieces for the body-box. These will consist of the front and back, $18\frac{1}{2}$ by 9 inches; the sides, 17 by 9; and the smaller pieces for the porch and entrance slides. To these add two pieces for the inner walls AA (Fig. 7), $17\frac{1}{2}$ by $8\frac{5}{8}$ inches, and $\frac{1}{2}$ inch thick. Place these inner walls at a distance of $1\frac{1}{4}$ inches from the front and back, leaving a space of exactly $14\frac{1}{2}$ inches between them. They should be grooved into the sides to a depth of $\frac{1}{4}$ inch on either side, finishing flush with the sides on the bottom edge, but ⅜ inch down at the top, to allow for the frame top bars.

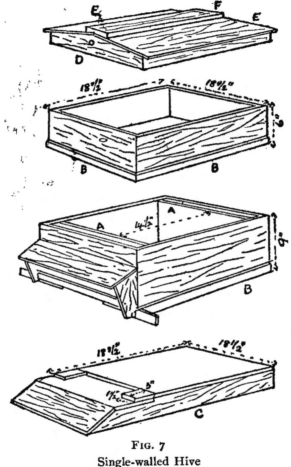

FIG. 7

Single-walled Hive

The space between the inner and outer walls must be filled in with strips of wood. Place these strips flush at the bottom, but ¼ inch below the level of the inner walls at top. It is well to bevel the top edges of these inside walls to a thickness of $\frac{1}{4}$ inch, and the bevel should be on the outer side. The porch and strip for entrance slides may be screwed on from the inside before sliding the inner walls into position. Half-inch wood will do for these. Rebate the slide strip on the inner bottom edge, and the slides on the outer top edge, but let them fit rather loosely. With regard to this, note that while hives require to be made correctly they should fit easily. Bees dislike nothing so much as the sharp jerks caused by the sectional parts of hives being tight.

The slides should be 10 inches long by $1\frac{1}{4}$ inches in width, while the plinths which run around the bottom edge at BB are 2 inches wide by $\frac{3}{4}$ inch thick. They are bevelled on the top edge as shown, and fall $\frac{1}{2}$ inch below the bottom edge of the body-box. They are slightly rebated on the inner bottom edge, in order to make an easy fit.

The floor-board consists of two pieces, 3 by $1\frac{1}{2}$ inches, C, on which are nailed $\frac{3}{4}$ inch boards. It will be noticed that it is necessary to sink the front board for the purpose of forming the entrance. This is known as a sunk entrance, and is far and away the best of its kind, as it entails no cutting of the body-box.

The lift consists of two pieces $18\frac{1}{2}$ by 6, and two others 17 by 6. It has plinths made and fitted the same as the ones on the body-box, except that they are on all four sides. For the roof two pieces $20\frac{1}{4}$ by 3 inches and two others

18¾ by 2 inches will be needed, the two former being shaped as shown at D. On this nail the two wide boards EE, which should overhang ¾ inch all round, and be capped by a ridge piece F. If any difficulty is experienced in obtaining these wide boards in deal, use American white wood for the purpose.

There are no plinths on the roof. It is made to fit right over the lift, and a stop is placed inside. This stop consists of strips ½ inch square, tacked all round at a distance of ¾ inch from the edge.

All corners of hives should be well screwed, the screw-heads being countersunk and the holes filled with putty. Bore an inch hole in the gable ends of roof for ventilation, and cover it on the inside with wire gauze. Three coats of paint may now be given, and the hive will be ready for use.

Shallow frame supers for this hive are made in exactly the same way as the body-box, but only 6 inches deep, and with a plinth all round. Two of these supers should be made for each hive. Frames should not be made but bought. They can be bought so cheaply that it is poor economy to attempt to make them.

A well-made hive of this description will last a lifetime if kept well painted, while a substantial saving will have been made in the initial cost.

INDEX